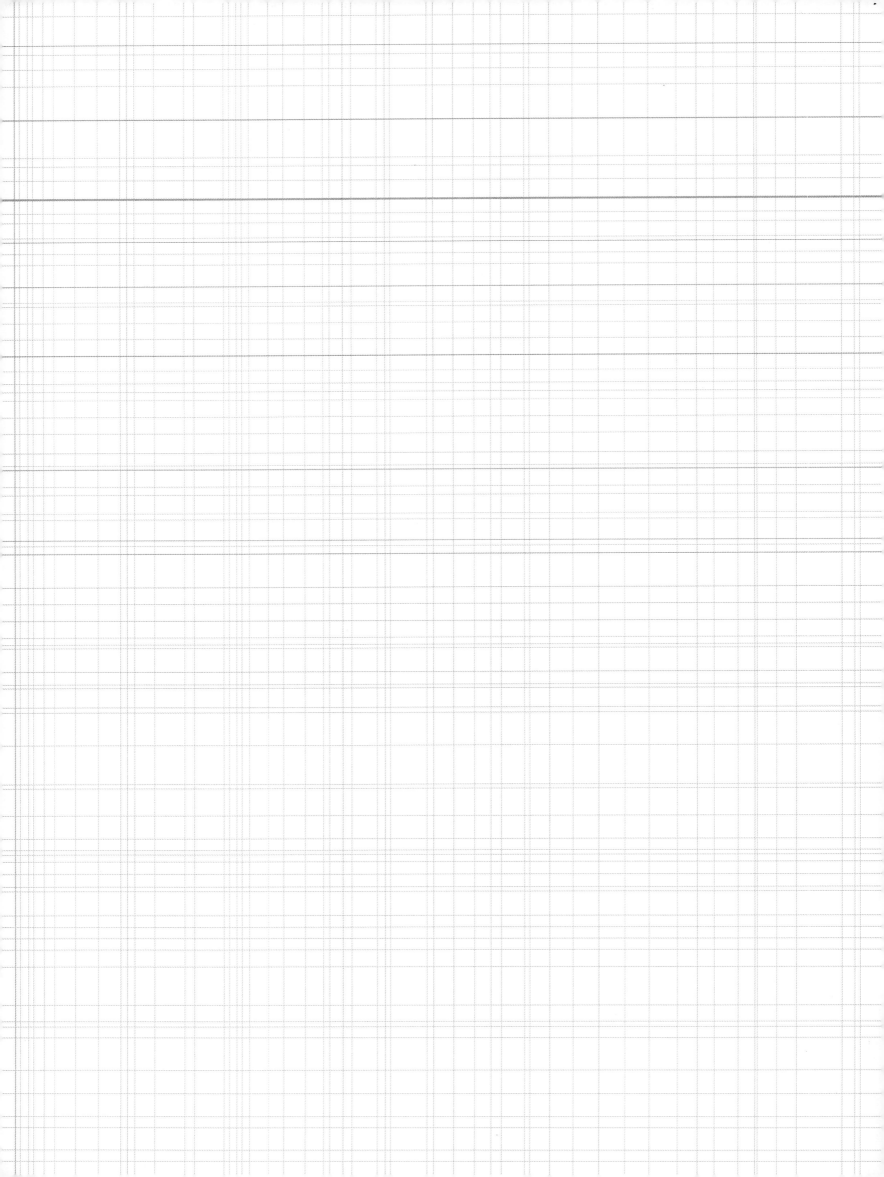

SURMOUNT
STYLE+COPY Ⅱ 突破风格与复制Ⅱ(下)

香港建筑科学出版社
唐艺设计资讯集团有限公司 策划

香港建筑科学出版社 编

天津大学出版社
TIANJIN UNIVERSITY PRESS

PREFACE

序言

This is a chaotic and complicated information age, in which people are overwhelmed by large amount of mixed information from all times around the world. Now the distinct line between time and space becomes hazy; classic and avant-garde turn to be short-lived. People get tired and lost under the fast changing of aesthetics, value, life style, and etc. If residential design is a lifestyle customization which will fix someone's living way, its key words should be how to balance the relationship among society, human and family.

Nowadays, China and many developing countries are in the high-speed developing period, referring to not only economy, but also rapid changing in value and aesthetics for growth of knowledge and culture. How can residential design not only serve local people, but also satisfy international needs, not only reflect the features of times, but also back to people-orientation? This requires designers pay more attention to society and broaden their knowledge. Residence design has exceeded its own meaning and value; it not only has fundamental and completed functions, but also need to have more added values in order to reflect the significance of the various levels, meet the changing trends and get close to the modern people on the present and spiritual needs for future living space.

Surmount Style+Copy has selected elaborately more than 80 outstanding works in major cities around the world designed by certain famous architects to fully demonstrate the tendency of global residence design. Architecture itself is an embodiment of thinking, which will show different meanings by different readers. We only have an analysis in basic aspects of façade, structure, ecology and sustainability, more deeper discoveries are waiting you to find in the book.

HongKong Architecture Science Press
Editorial Board

这一个混沌而繁杂的信息化时代，人们被穿越古今中外、形形色色的资讯所淹没，时间和空间变得朦胧，经典与前卫只是昙花一现。审美观、价值观、生活方式等快速地变化着，人们变得疲惫而迷失。如果住宅设计是一种生活方式的订制，如何平衡社会、人、家三者之间的关系则是住宅设计的焦点问题。

中国和很多发展中国家都处于高速发展期，不仅仅是经济的增长，知识与文化的增长也给价值观、审美观带来更快速的变化。如何做到既要服务当地群众，又要满足国际需求，既反映时代特色，又回归以人为本的本源？这需要建筑师更加关注社会、具备更宽广的知识层面。住宅设计已经完全超越其本身的意义和价值，它不仅要具有基础和完备的功能，还需要具备更多的附加价值，体现各个层面错综复杂的意义，更要面对不断变化的趋势，符合现代居住者以及对未来居住空间的精神需求。

《突破风格与复制》特意从世界著名建筑事务所中遴选全球主要城市的 80 多个精彩案例，是为了更加多元、全面地展示世界住宅设计的趋势。建筑本身就是一种思想的展示，不同的读者可以赋予它不同的寓意，我们仅从立面、结构、生态、可持续发展等基本方面加以分析，更多的、更深层的隐喻期待读者去发掘、去解读。

香港建筑科学出版社
编委会

CONTENT

目录

SHAPE 造型

BALCONY 阳台

MATERIAL 材料

SUSTAINABILITY 可持续性

Jujuy Redux, Rosario, Argentina

阿根廷罗萨里奥胡胡伊 Redux 公寓

Architect: Patterns + MSA
Principals: Marcelo Spina, Georgina Huljich, Maximiliano Spina
Location: Rosario, Argentina
Site Area: 220 m²
Gross Floor Area: 1,350 m²
Floors: 9
Height: 28.5 m
Photography: Gustavo Frittegotto

设计公司：Patterns + MSA
主设计师：Marcelo Spina、Georgina Huljich、Maximiliano Spina
地点：阿根廷罗萨里奥市
占地面积：220 平方米
建筑面积：1 350 平方米
层数：9
高度：28.5 米
摄影：Gustavo Frittegotto

Jujuy Redux is a mid-rise apartment building located in Rosario, Argentina. As architects' second commission for an apartment building in Argentina (the first one located also on Jujuy Street), the project presented the possibility of rethinking urban housing. The project consists of thirteen small, shared-floor units and a duplex organized in a cross-ventilated layout. The first-floor provides parking for 10 vehicles, and a common terrace on the 8th floor provides outdoor leisure spaces.

胡胡伊 Redux 公寓坐落于阿根廷罗萨里奥市，是一栋中高层公寓建筑。作为设计师在阿根廷的第二个受委托的公寓项目（第一个项目也坐落在胡胡伊街），本案为世人呈现了重新思考城市住房的可能性。项目包括 13 个小型的共享单元和一间采取交叉通风布局的复式套房。首层提供了 10 个停车位，8 楼设置了公共露台作为户外休闲空间。

Occupying a corner lot, the new development aims to revitalize Rosario's traditional neighborhood of "Pichincha". Adjacent to both the historic downtown and the Parana Riverfront, it enjoys generous street sizes, large amounts of vegetation and a low urban density of cultural heritage. It is an ideal environment for young families and students. As a result, the demand for new mid-rise interventions stands at an all-time high.

Jujuy Redux proposes a subtle delineated mass, operating both at the scale of the entire volume and the scale of each apartment. This flexible duality overcomes issues that exist in many mid-rise housing typologies, such as the occurrence of fixed, scalar transformations that play either with the envelope detached from the units, or with the units alone.

A transition from mass to volume, from volume to surface, induces a visual and physical distortion at the pedestrian level. More importantly, it enables the weighty appearance of the building to sinuously dematerialize towards the corner, allowing the social space of each apartment to visually connect with the pedestrian activities on the street level.

新公寓占据了一片位于角落的场地，以振兴罗萨里奥的传统街区 "皮钦查" 为目标。公寓邻近历史悠久的市中心和巴拉那河岸，周边街道宽阔有序、绿树成荫且文化遗产密度低，是适合年轻家庭和学生居住的理想环境。所以此处对中高层公寓的需求旺盛。

胡胡伊 Redux 公寓呈现了一个微妙的体块，同时兼顾到了公寓整体体量和每间公寓单元的大小。这种灵活的双重性可以解决许多中高层住宅中普遍存在的问题，如固定标量转换的发生，它们或对于单元分隔的外壳起作用，或对独立单元产生影响。

整体到体量、体量到立面的过渡在人行道层面产生了视觉上和物理上的变形。更重要的是，这种变形能让建筑看似沉重的外形向角落弯曲，直至消失，同时使各单元的社交空间和街道上的行人活动在视觉上联系在一起。

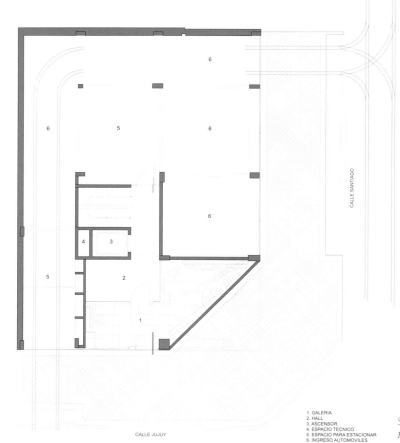

1. GALERIA
2. HALL
3. ASCENSOR
4. ESPACIO TECNICO
5. ESPACIO PARA ESTACIONAR
6. INGRESO AUTOMOVILES

CALLE JUJUY

CALLE SANTIAGO

Site Plan
总平面图

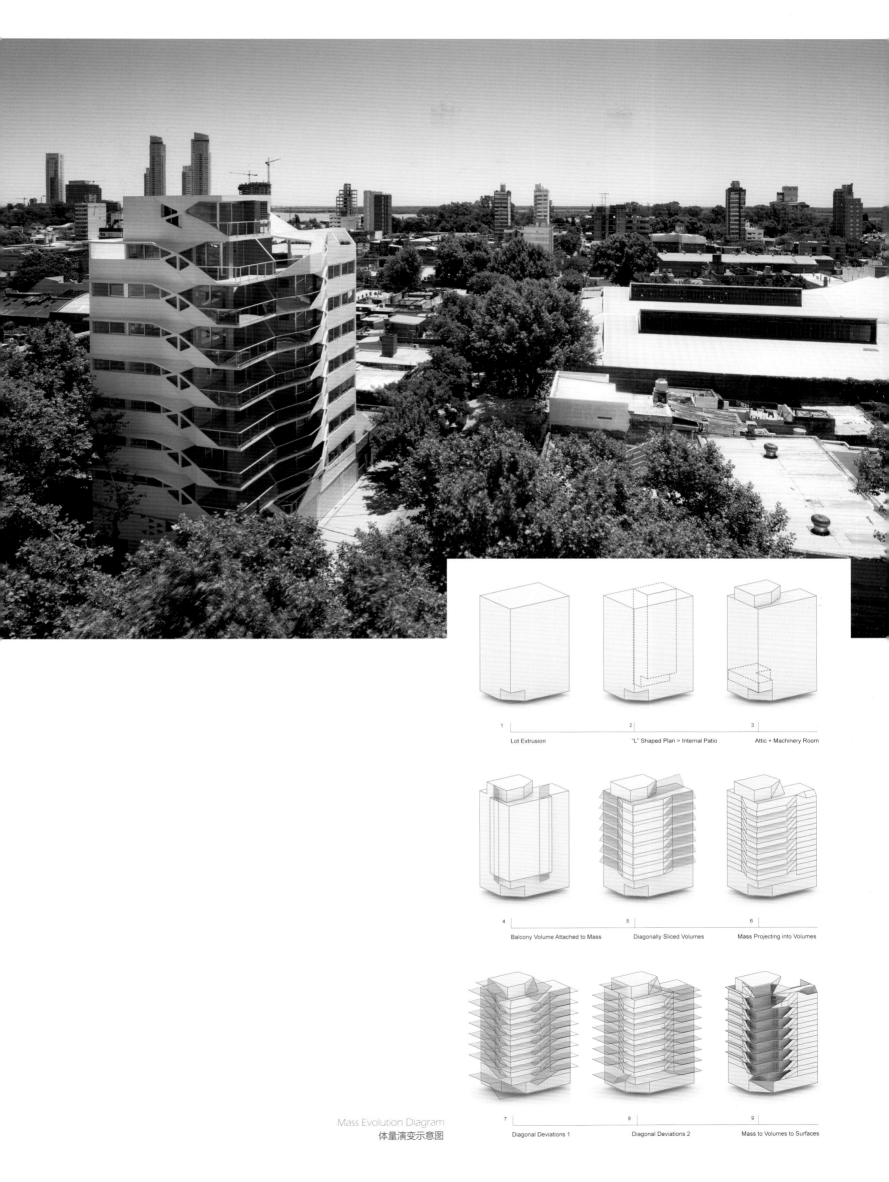

Mass Evolution Diagram
体量演变示意图

1	2	3
Lot Extrusion	"L" Shaped Plan > Internal Patio	Attic + Machinery Room

4	5	6
Balcony Volume Attached to Mass	Diagonally Sliced Volumes	Mass Projecting into Volumes

7	8	9
Diagonal Deviations 1	Diagonal Deviations 2	Mass to Volumes to Surfaces

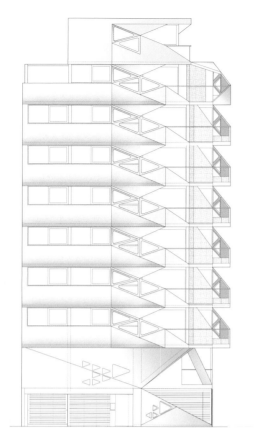

South Elevation
南立面图

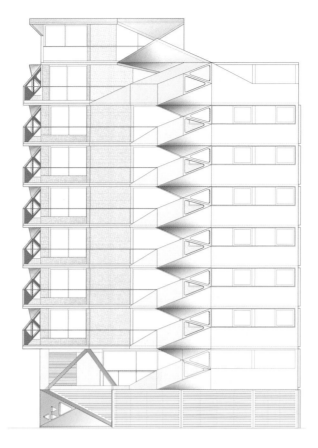

East Elevation
东立面图

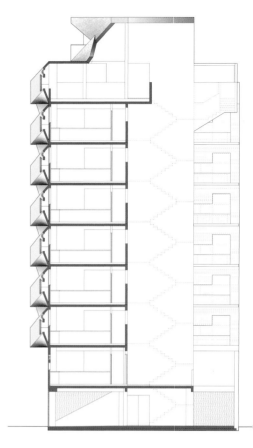

Section
剖面图

The project is built almost entirely of cast-in-place concrete, and the interior space of the balconies is clad with grey venetian tile. This arrangement produces a sense of contrast between the outer white concrete skin and the inner walls of the balcony, while also accentuating a sense of spatial enclosure within the outdoor living space of each apartment.

Jujuy Redux's balconies are conceived as highly articulated pieces that incorporate apertures, railing, direct and indirect LED lighting as well as material changes.

Problematically, balconies are the important cultural element in mid-rise residential buildings of South America. Typologically, balconies have become the playground for formalism, often neglecting issues of spatial integration into an overall scheme, and worse, dissociating them from social issues and human inhabitation. By contrast, the formal, spatial and material treatment of the balconies of Jujuy Redux is one of the most significant, innovative and socially performative aspects of the project.

A built-in bench provides a place for leisure activities and small gatherings, while allowing inhabitants to tailor its use according to their individual needs.

本项目几近是由浇注混凝土建造，阳台的内部空间覆盖了灰色百叶瓦。这种搭配能在白色混凝土表皮和阳台内壁之间产生反差，同时也加强了各单元的户外生活空间的空间围合感。

胡胡伊 Redux 公寓的阳台就像高度铰接的部件，将开口、栏杆、直接与间接 LED 照明和不同材料相互配合起来。

问题在于，阳台在南美洲中高层建筑中是不可忽视的文化元素。其中最为常见的是阳台成为了形式主义的操场，往往忽略了将空间整合纳入整体策划内；更糟的是，脱离了社会问题和人类居住需求。相比之下，胡胡伊 Redux 公寓可称得上是最为突出、创新和富有社交表达性的项目，尤其表现在阳台的形式、空间和材料处理方面。

内置的长椅为居民提供了休闲活动和小型聚会的场所，同时可以让居民根据自己的个性化需求来改变其使用方式。

FEATURE 特点分析

BALCONY

We recognize the open cultural conditions associated with balconies in housing buildings that are not yet coded with an architectural typology. Jujuy Redux's balcony design takes inspiration from the polygonal bay window as well as from the horizontal balcony. While traditional bay windows perform strictly as interior spaces, Jujuy Redux balconies perform directionally as they open up towards the corner, allowing for both exposed spaces with oblique vistas as well as more intimate ones, sheltered from the weather and direct views from the street.

阳台

设计师认识到，开放的文化条件和住宅建筑的阳台并未真正在建筑类型学中统合在一起。胡胡伊 Redux 公寓的设计灵感来自多边形的凸窗和水平阳台。传统的凸窗被严格应用于内部空间中，胡胡伊 Redux 公寓的阳台直接向角落敞开，让外部空间、远景和近景亲密接触，并遮挡风雨和从街上直视而来的视线。

Following the geometry of the balcony system, the triangular openings along the shell open up a series of threshold spaces to control sunlight, natural ventilation and views. Where the shell doubles up, it is perforated, creating a passive solar technique that, in addition, the cross-ventilated layout of the apartments helps produce an effective natural cooling system. Similarly, the chamfered corner and main entry are punctured in a larger number, allowing the building porch to receive filtered morning light while offering passers-by voyeuristic peeks into the building.

White walls have traditionally been associated with the details and ornamentation, often resulting in a stale materiality, like the overt whitewash found in many nineteenth-century buildings. Alternatively, Jujuy Redux explores a different kind of whiteness: one that privileges overall plasticity over local materiality and engages in high or low contrast to address depth and flatness, subtle white to off-white.

配合阳台的几何形状，沿着外壳布置的三角开口打开了一系列的临界空间，用于控制光线、自然通风和视野。在外壳折叠的地方进行穿孔，创造了一种被动式太阳能技术；此外，公寓交叉通风的布局有助于形成一个有效的自然冷却系统。同样，在倒角和主入口处也采用了大量的穿孔布局，有利于建筑的走廊能沐浴到经过过滤的晨光，路人经过时也可以由此窥到建筑里的构造。

传统的白墙一般会和细节与装饰联系在一起，这样会产生陈旧的感觉，就像许多 19 世纪建筑中使用的纯白色粉刷的墙。因此，胡胡伊 Redux 尝试了另一种不同的白色：介于精白和灰白之间，是在本地材料中具有极强可塑性、能产生高或低对比度来强调出深度和平整度的色彩。

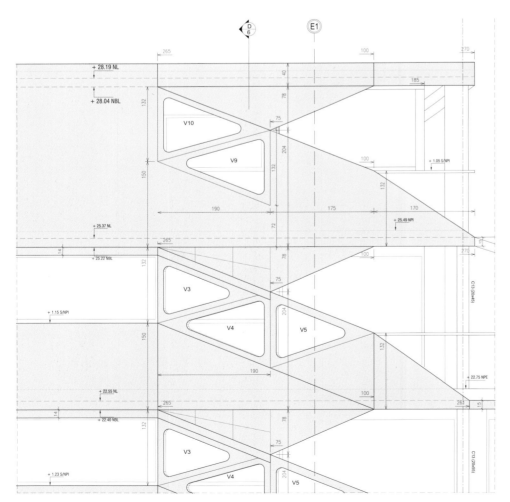

South Elevation Detail
南立面细节图

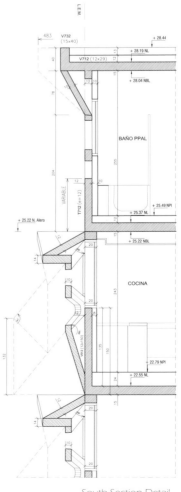

South Section Detail
南面剖面细节图

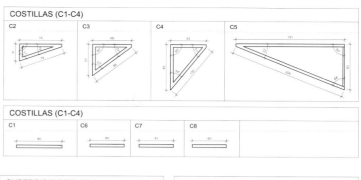

COSTILLAS (C1-C4)

C2 C3 C4 C5

COSTILLAS (C1-C4)

C1 C6 C7 C8

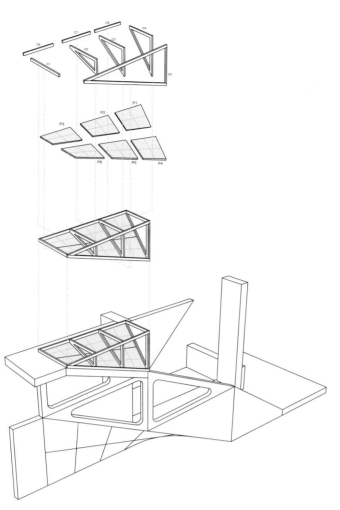

Balcony Detail 1
阳台细节图 1

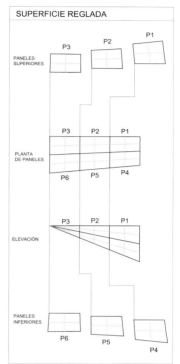

SUPERFICIE REGLADA

PANELES
SUPERIORES

P3 P2 P1

PLANTA
DE PANELES

P3 P2 P1
P6 P5 P4

ELEVACIÓN

P3 P2 P1

PANELES
INFERIORES

P6 P5 P4

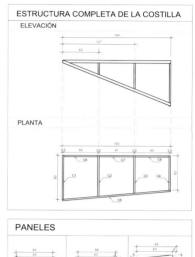

ESTRUCTURA COMPLETA DE LA COSTILLA

ELEVACIÓN

PLANTA

PANELES

P3 P2 P1

P6 P5 P4

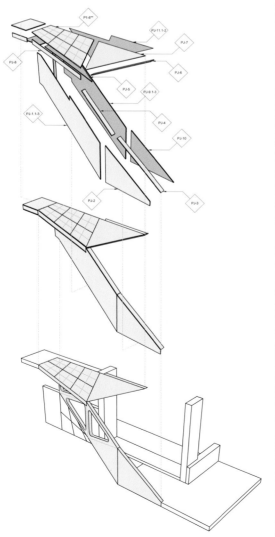

Balcony Detail 2
阳台细节图 2

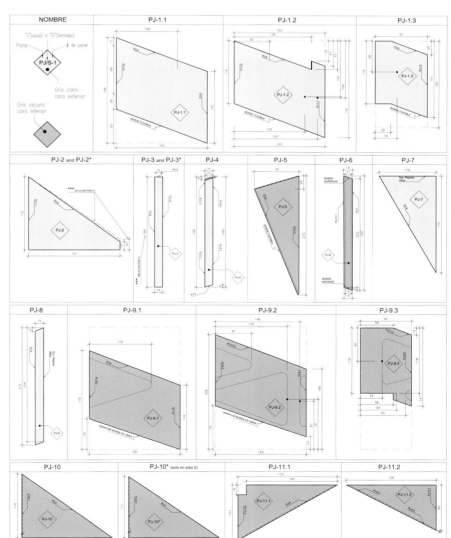

NOMBRE PJ-1.1 PJ-1.2 PJ-1.3

"J"(Jujuy) o "S"(Santiago)
Panel — de panel
PJ/S-1

Gris claro:
cara exterior

Gris oscura:
cara interior

PJ-2 and PJ-2* PJ-3 and PJ-3* PJ-4 PJ-5 PJ-6 PJ-7

PJ-8 PJ-9.1 PJ-9.2 PJ-9.3

PJ-10 PJ-10* (solo en piso 2) PJ-11.1 PJ-11.2

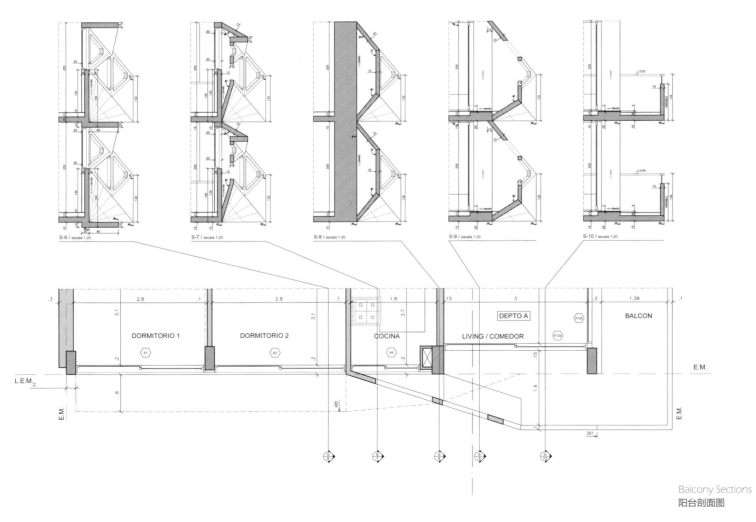

S-6 / escala 1:20   S-7 / escala 1:20   S-8 / escala 1:20   S-9 / escala 1:20   S-10 / escala 1:20

DORMITORIO 1   DORMITORIO 2   COCINA   DEPTO A   BALCON

LIVING / COMEDOR

E.M.

L.E.M.

E.M.

Balcony Sections
阳台剖面图

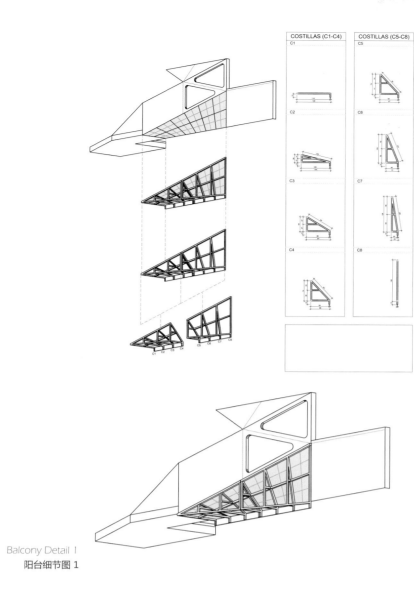

Balcony Detail 1
阳台细节图 1

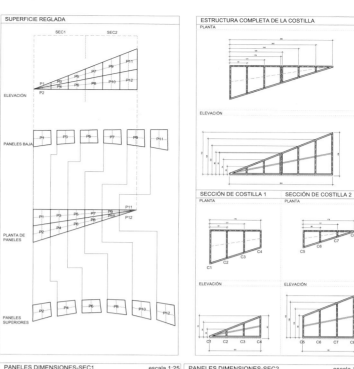

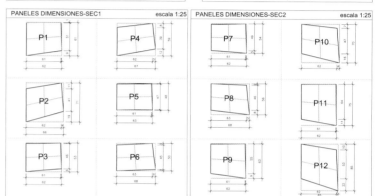

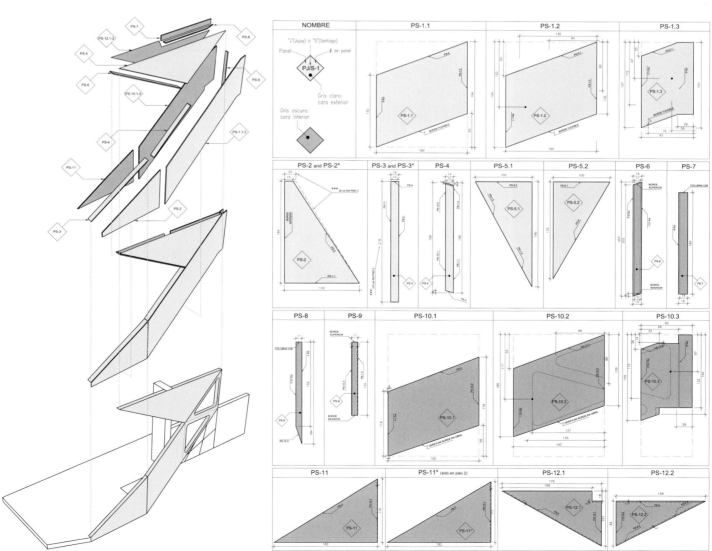

Balcony Detail 2
阳台细节图 2

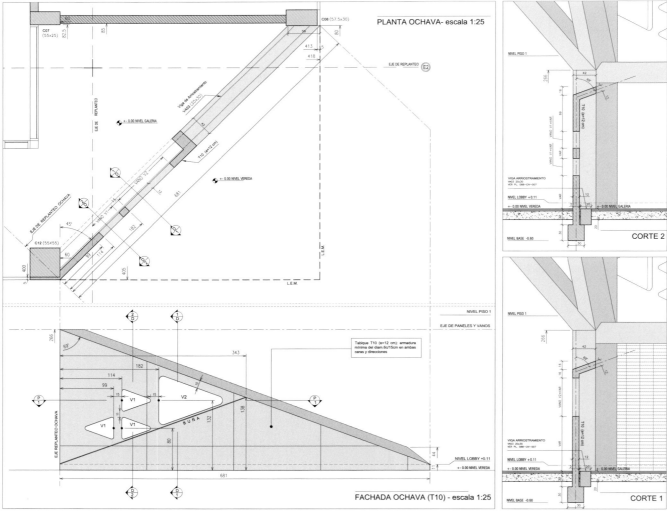

PLANTA OCHAVA- escala 1:25

FACHADA OCHAVA (T10) - escala 1:25

CORTE 2

CORTE 1

Entrance Detail
入口细节图

At first-floor, concrete cross braces receive the diagonal deviations produced by the balconies, creating a double-height urban corner free of columns. This cantilevered corner gives way to an inconspicuous building entrance located at Jujuy Street, followed by a sequence of spaces: a gated porch and main hall linked by a two-storey glazed doorway, and the elevator lobby, all clad in polished Carrara marble.

在首层，混凝土肋板受到阳台对角偏差的影响，形成了双高无柱的城市一角。带悬臂的拐角为位于胡胡伊街不起眼的建筑入口让出了道路，并带来一系列的空间变化：一道带有门控的门，由两层高釉面门道连接的主厅以及电梯厅都覆盖了一层抛光的卡拉拉大理石的外衣。

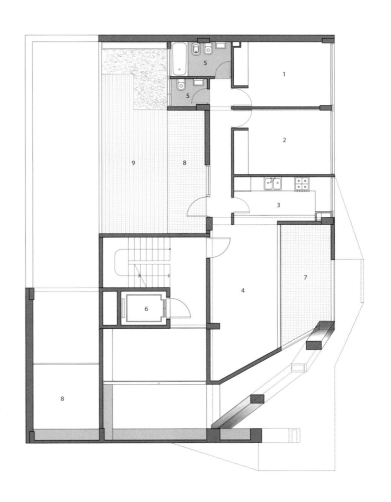

1. DORMITORIO 1   6. ASC
2. DORMITORIO 2   7. BALCON
3. COCINA   8. GALERIA
4. LIVING / COMEDOR   9. TERRAZA
5. BAÑO

1st Floor Plan
首层平面图

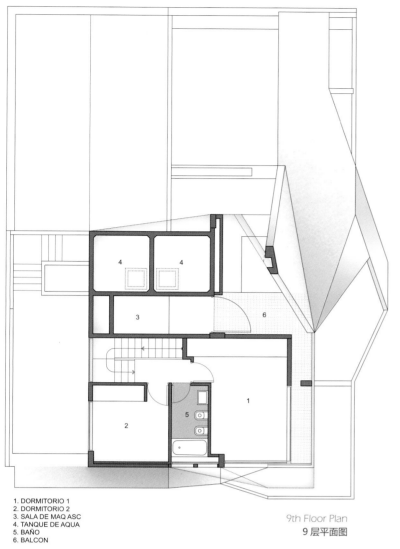

1. DORMITORIO 1
2. DORMITORIO 2
3. SALA DE MAQ ASC
4. TANQUE DE AQUA
5. BAÑO
6. BALCON

9th Floor Plan
9 层平面图

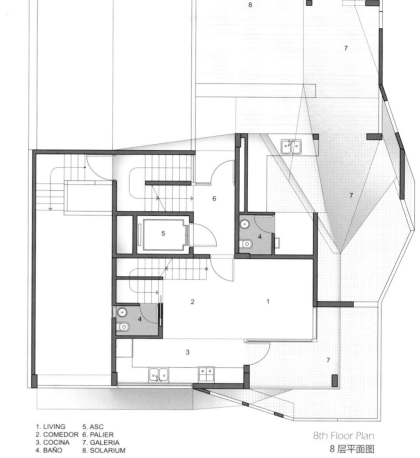

1. LIVING   5. ASC
2. COMEDOR   6. PALIER
3. COCINA   7. GALERIA
4. BAÑO   8. SOLARIUM

8th Floor Plan
8 层平面图

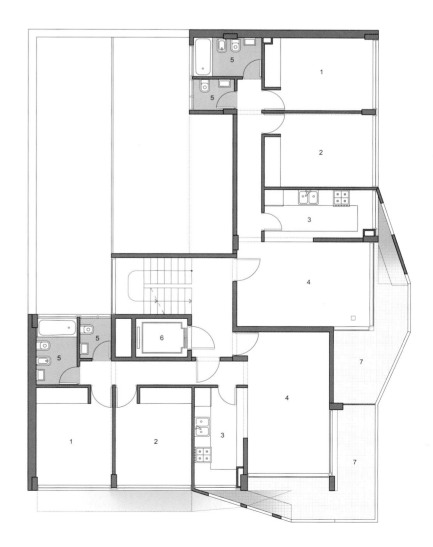

1. DORMITORIO 1
2. DORMITORIO 2
3. COCINA
4. LIVING / COMEDOR
5. BAÑO
6. ASC
7. BALCON

2nd-6th Floor Plan
2~6 层平面图

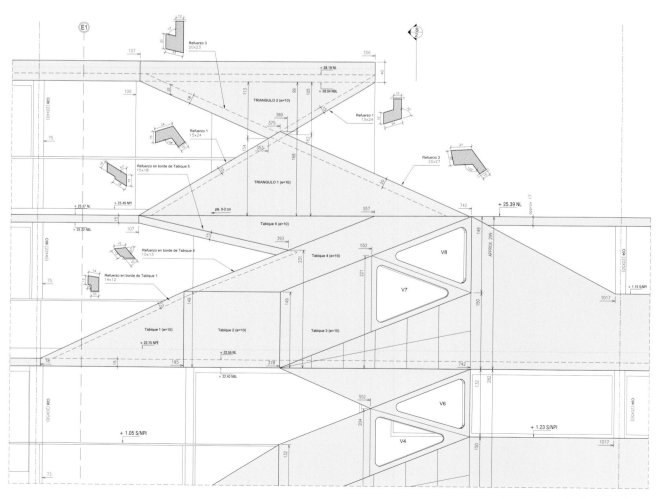

The formal sequence of diagonal deviations intensifies at the roof level, where the building tapers to accommodate a set of mechanical spaces, a corner duplex, a common terrace with a solarium and semi-covered areas for barbecuing — all of them enjoying attractive views of downtown and the riverfront. Visually, the building opens up and lightens its contorted mass as it rises above the ancient grove in sequenced diagonal recesses to meet the city skyline.

对角偏差的序列组合让屋顶更为特别，公寓在屋顶处逐渐缩小，以为机械设备、角落的复式套房、设置有日光浴场和半露天烧烤区的公共露台提供空间，如此一来，这些地方都能尽览市中心和河岸的无限风光。在视觉上，公寓打开并减轻了其扭曲的体量，依次缩进的对角结构高出古树林，和城市的天际线进行亲密接触。

Terraza Elevation
露台立面图

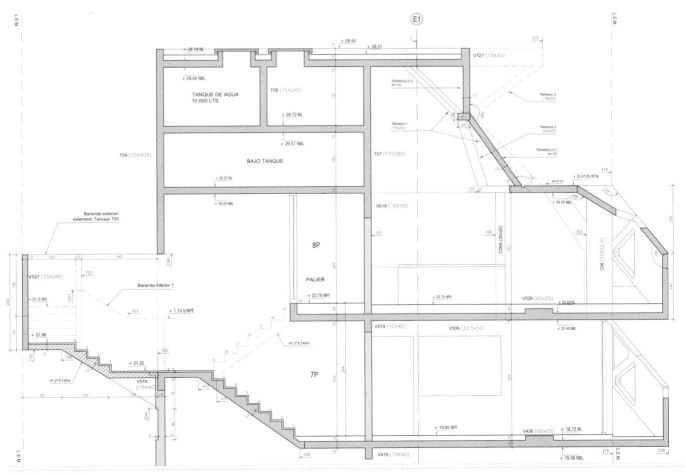

Terraza Section
露台剖面图

FAÇADE 立面
BALCONY 阳台
STRUCTURE 结构
OPEN SPACE 开放空间

Westerdok Apartment Building, Amsterdam, the Netherlands

荷兰阿姆斯特丹韦斯特多克公寓楼

Architect: MVRDV
Client: O.M.A. (Ontwikkelings Maatschappij Apeldoorn)
Location: Amsterdam, the Netherlands
Site Area: 730 m²
Gross Floor Area: 6,000 m²
Floors: 11

设计公司：MVRDV
客户：O.M.A. (Ontwikkelings Maatschappij Apeldoorn)
地点：荷兰阿姆斯特丹市
占地面积：730 平方米
建筑面积：6 000 平方米
层数：11

Openness as design concept: the minimum amount of materials used, glass, steel and concrete, achieves maximum openness for the façade.

开放的设计理念：使用最少的材料，用玻璃、钢材、混凝土让立面达到最大的开放度。

The building with a total surface of 6,000 m² contains 46 apartments and a day-care center. Each apartment has a balcony of varying depths which stretches as bands along the entire façade, offering varied outside spaces and views over the western docklands of Amsterdam. The floor-to-ceiling glass façade can be fully opened and contrasts with the other buildings within the so-called "VOC Cour" port redevelopment that are mainly made of brick.

The urban plan is a closed city block with buildings of differing heights surrounding a central court. After two earlier urban plans failed, the client O.M.A. (Ontwikkelings Maatschappij Apeldoorn) has in fact determined the current urban plan. The MVRDV building is located inside the court with one façade facing the waterfront of the Westerdok.

这座公寓面积为 6 000 平方米，包括 46 间公寓和一个日托中心。每个公寓有一个阳台，它们有不同的进深，像带子一样沿整个立面展开，提供了不同的外部空间和阿姆斯特丹西部港区的景观。落地玻璃立面完全对外开放，这与主要为砖结构的 VOC Cour 港重建区内的其他建筑形成对比。

在城市设计中，该地是围绕着中央广场的一系列不同高度的建筑群。在两个早期的城市规划失败以后，客户 O.M.A. (Ontwikkelings Maatschappij Apeldoorn) 事实上使用了现代城市设计手法。本案位于广场内部，一个立面朝向韦斯特多克的滨水地带。

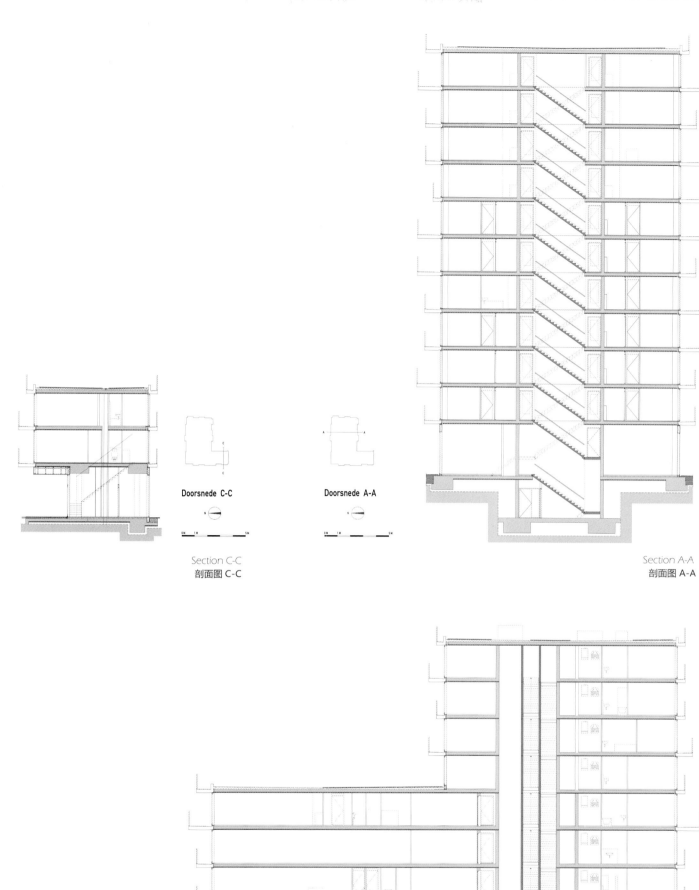

Doorsnede C-C

Doorsnede A-A

Section C-C
剖面图 C-C

Section A-A
剖面图 A-A

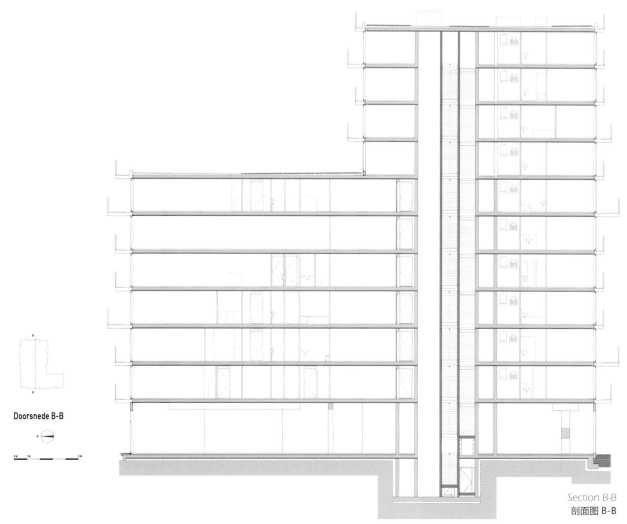

Doorsnede B-B

Section B-B
剖面图 B-B

FEATURE 特点分析

BALCONY

Building's envelopes(façade, balcony) are transparent, maximize the exposure and lighting inside the building. The balconies with less guardrails make the floor outstanding, and form contrast with the concise glass façade, which is highly decorative. Designing the complex (balcony, window) residential façades into modern and minimalist office façades (glass curtain wall, shield) optimizies façade features, while making the building look more attractive.

阳台

建筑的围护结构（立面、阳台）被透明化，使建筑内部视野和采光最大化；淡化了护栏的阳台使底板格外醒目，与简洁的玻璃立面形成对比，具有很强的装饰性。把复杂的（阳台、窗）住宅立面设计成现代、简约的写字楼立面（玻璃幕墙、遮阳板），在优化立面功能的同时使建筑外观更具魅力。

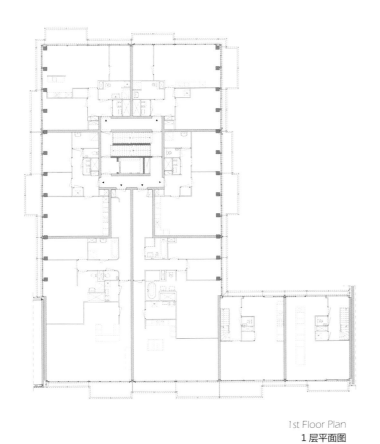

1st Floor Plan
1层平面图

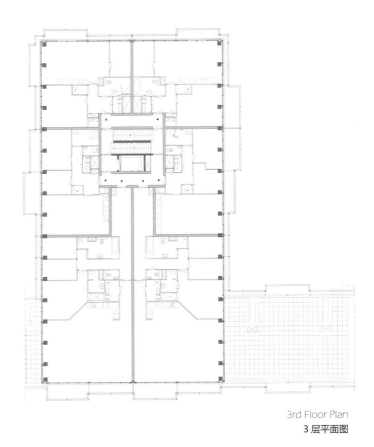

3rd Floor Plan
3层平面图

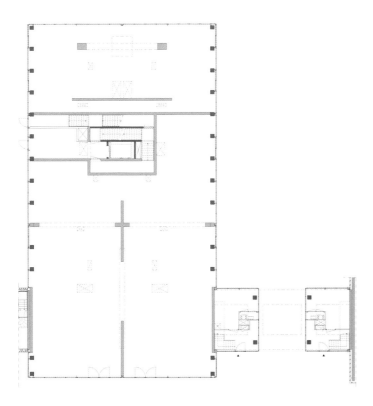

Underground Floor Plan
地下一层平面图

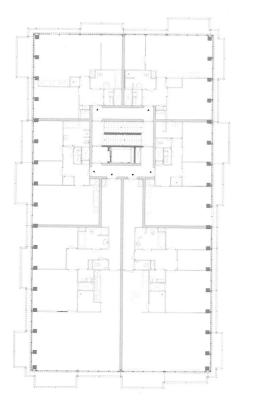

4th Floor Plan
4 层平面图

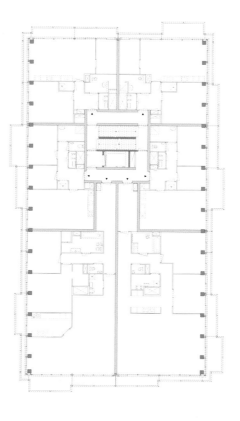

5th Floor Plan
5 层平面图

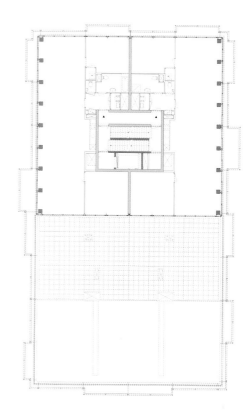

7th Floor Plan
7 层平面图

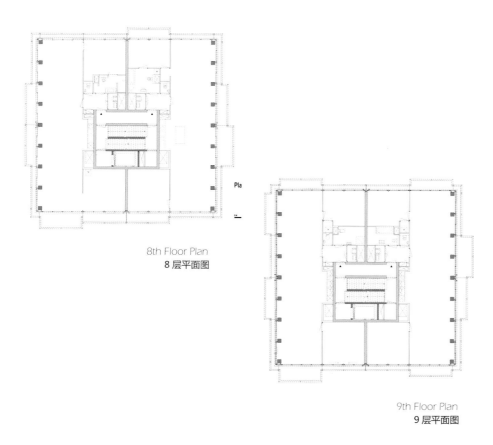

8th Floor Plan
8 层平面图

Pla

9th Floor Plan
9 层平面图

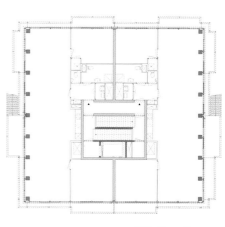

10th Floor Plan
10 层平面图

VIEW 视野

FAÇADE 立面

WINDOW 窗

STRUCTURE 结构

Schlump ONE, Hamburg, Germany
德国汉堡 Schlump ONE 大厦

Architect: J. Mayer H. Architects
Client: Cogiton, Projekt Eimsbuettel GmbH
Location: Hamburg, Germany
Photography: Jan Bitter, Ludger Paffrath

设计公司：J. Mayer H. Architects
客户：Cogiton、Projekt Eimsbuettel GmbH
地点：德国汉堡市
摄影师：Jan Bitter、Ludger Paffrath

The project "Schlump ONE" is located directly at the underground station Schlump in Eimsbüttel District in Hamburg. The original administration building was built in the 1950s and was gutted, renovated and expanded in the 1990s, and has now been converted into an office building with four units per floor for rent. The existing Data Processing Center in the courtyard has been transformed into a private university and a new building is built.

Schlump ONE 大厦位于德国汉堡市 Eimsbüttel 区的 Schlump 地铁站上。原行政大楼建于 20 世纪 50 年代，于 20 世纪 90 年代被拆建、修复、扩建，现在已经被改造成一座办公大楼，每层有四个可以出租的单元。在庭院里现有的数据处理中心已被改造成一所私立大学，并扩建了一座新的建筑。

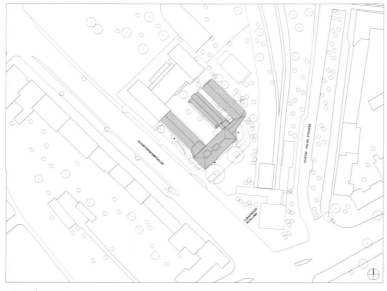

FEATURE 特点分析

FAÇADE

Combined with the extension, the building has a U-shaped plan that stretches around a courtyard at the rear. The curved forms are continued inside the building, where partitions have rounded openings that form façades of the corridors.

立面

结合扩建部分，建筑被设计成 U 形，在后方围绕着庭院展开。曲线设计在建筑内部得到延续，不同的区域以圆形开口划分，并形成了走廊的立面。

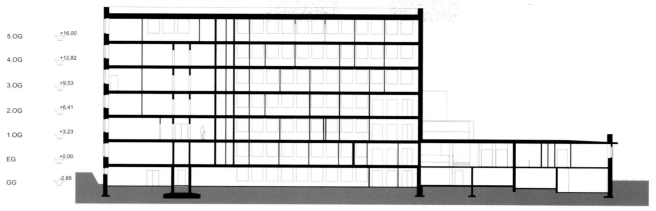

5.OG	+16,00
4.OG	+12,82
3.OG	+9,53
2.OG	+6,41
1.OG	+3,23
EG	+0,00
GG	-2,85

Section A-A
剖面图 A-A

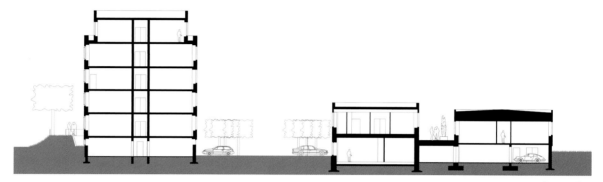

Section B-B
剖面图 B-B

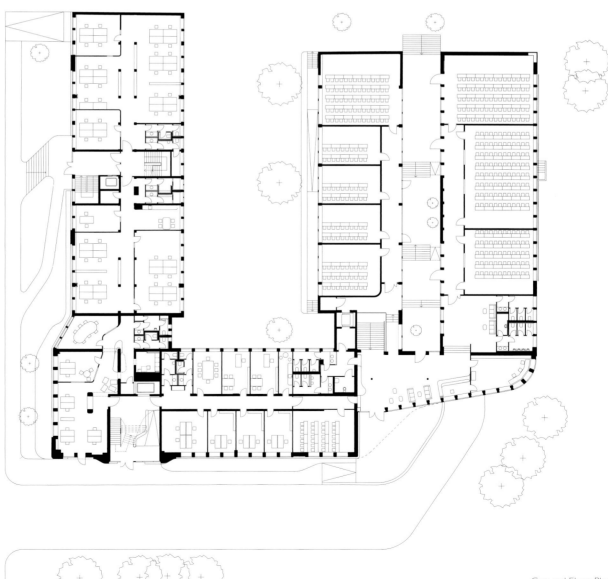

Ground Floor Plan
首层平面图

The gridded façade of the 1950s office building is replaced with organically curved glass and white render.

The building's façade has been completely renovated and redesigned to form a single unit that freely interprets the original building's 1950s linear design.

"We tried to design a façade that would be a bit more free, something less strict and linear" explained Wilko Hoffmann of J. Mayer H.

有机设计的曲面玻璃和白色墙体取代了 20 世纪 50 年代办公楼的网格立面。

建筑的外立面已被完全翻新和重新设计，以形成一个独立的单元，自由地演绎了原有建筑在 20 世纪 50 年代时的线性设计。

"我们试图赋予建筑立面一种更自由的感觉，少些僵化和线性感。" J. Mayer H. 公司的设计师威尔科·霍夫曼说道。

The organic formal language of the façade is continued in the design of interiors. The project is embedded in a sophisticated, open space planning design with oversized tree sculptures.

立面有机的形式语言在室内得到了延续。项目采用复杂的开放空间规划设计，并设计成了超大的树形雕塑形式。

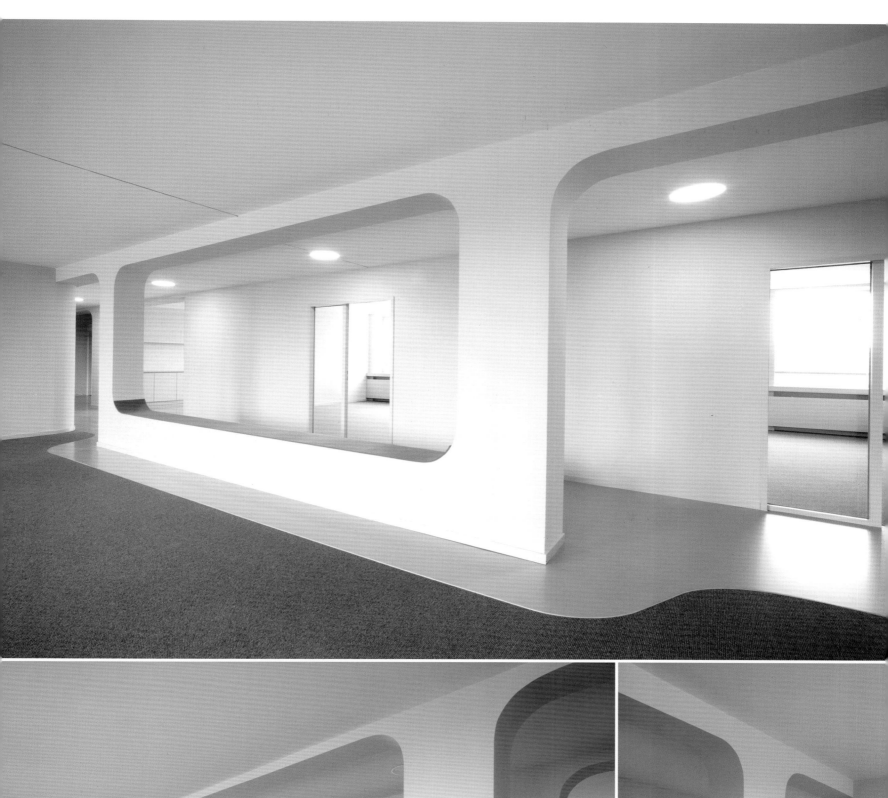
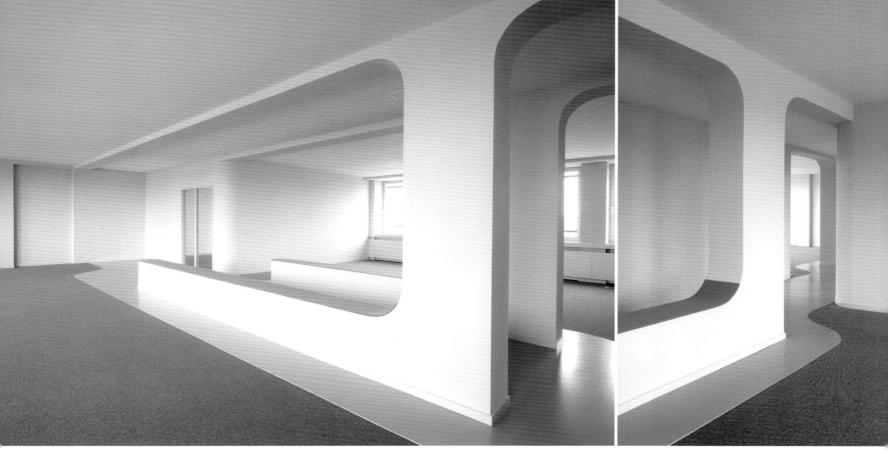

MATERIAL 材料

COLOR 色彩

BALCONY 阳台

FAÇADE 立面

10 Weymouth Street, London, UK

英国伦敦韦茅斯街 10 号

Architect: *Make Architects*
Client: Ridgeford Properties Ltd.
Location: London, UK
Site Area: 1,580 m²
Floors: 7
Photography: Z Olsen, Make Architects

设计公司：Make Architects
客户：Ridgeford 房地产公司
地点：英国伦敦市
占地面积：1 580 平方米
层数：7
摄影：Z Olsen、Make Architects

A relatively undistinguished 1960s block in Fitzrovia has been transformed by this highly distinctive refurbishment scheme, which increases the office and residential accommodation and gives the building a striking new identity.

坐落在 Fitzrovia 区的一栋建于 20 世纪 60 年代的平庸建筑依据与众不同的翻新方案被加以改建，并增加了办公和住宅功能，赋予该建筑一个引人注目的新身份。

Using imaginative and sensitive design, the resultant scheme has created 12 new luxury apartments by extending the rear elevation, three new penthouse apartments on the existing penthouse level and an extra floor which contains the fourth penthouse.

The existing building presented two very different façades to the surrounding area. In response to these contrasting conditions, a highly modelled brass-clad elevation was created at the rear of the building, which greatly enhances the appearance of this fascinating pocket of urban space. Each apartment features a projecting balcony with a perforated brass screen which mirrors the façade pattern and blends harmoniously with the neighboring buildings.

By contrast, the Weymouth Street elevation features a discreet structure addition to the existing structure, which is set back and clad in a darker alloy of the same brass. The unique appearance of the brass cladding will gradually evolve over time due to the oxidation process.

设计师的设计极具感性和想象力，通过扩展建筑后部的结构，增加了 12 套新豪华公寓，还在原来复式套房的楼层上增加了 3 间复式套房，最后又额外加设了一层楼，安置第 4 间复式套房。

原来的建筑的两个截然不同的立面，朝向周边地区。为了响应这些对比鲜明的条件，建筑的后部立面覆有黄铜，极大地提升了建筑在这个城市空间的迷人魅力。每间公寓都有一个突出的阳台，镂空黄铜面板也成为了立面图案的一部分，和周围建筑和谐地融为一体。

相比之下，朝向韦茅斯街的原有立面并不显眼，现在则被改造成缩进结构，同时也覆盖上颜色较暗的黄铜外皮。这层独特的黄铜外皮随着时间的推移将逐渐被氧化。

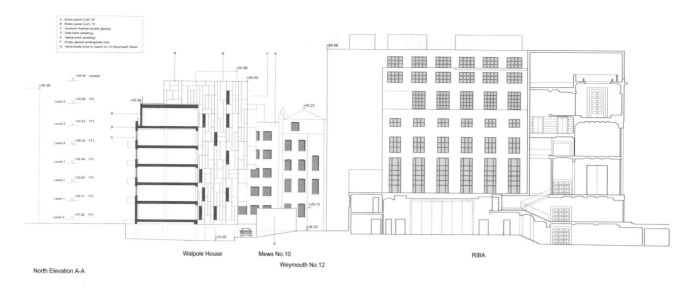

A Brass panel CuZn 30
B Brass panel CuZn 15
C Aluminium framed double glazing
D Dark brick (existing)
E Yellow brick (existing)
F Single glazed wintergarden box
G Hand-made brick to match no.12 Weymouth Street

+48.08

+49.49 parapet

Level 6   +45.88 FFL
           +45.88
+50.99
+49.49
+45.22

Level 5   +42.43 FFL
A
D
Level 4   +39.38 FFL
C
Level 3   +36.49 FFL
Level 2   +33.60 FFL
+29.75
Level 1   +30.71 FFL

Level G   +27.82 FFL
+26.20
+24.60

G

+54.58

Walpole House   Mews No.10   RIBA

North Elevation A-A   Weymouth No.12

+52.47

+49.95
C   F   B
+49.49
+50.99
B

+49.49 parapet
+45.88
Level 6   +45.88 FFL
+44.29

Level 5   +42.43 FFL
Level 4   +39.38 FFL
C
Level 3   +36.49 FFL
Level 2   +33.60 FFL
Level 1   +30.71 FFL
Level G   +27.82 FFL
+27.26

+24.60

Hallam Court   Walpole House

West Elevation B-B

1:100   0   10m   20m

Elevation 1
立面图 1

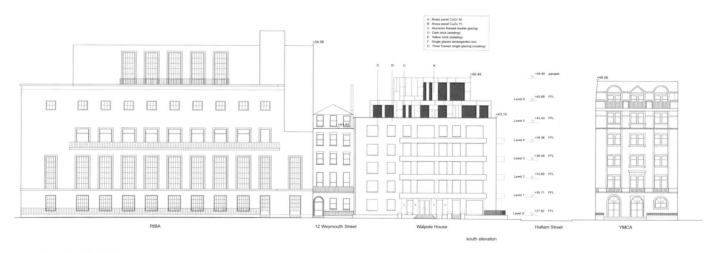

A Brass panel CuZn 30
B Brass panel CuZn 15
C Aluminium framed double glazing
D Dark brick (existing)
E Yellow brick (existing)
F Single glazed wintergarden box
G Timer framed single glazing (existing)

+54.58

C D C   A
+49.49
+49.49 parapet
+49.06

Level 6   +45.88 FFL
+43.16
+41.43

Level 5   +42.43 FFL
Level 4   +39.38 FFL
Level 3   +36.49 FFL
Level 2   +33.60 FFL
Level 1   +30.71 FFL
Level G   +27.82 FFL

RIBA   12 Weymouth Street   Walpole House   Hallam Street   YMCA

south elevation

Proposed South Elevation A-A

parapet +49.49
+45.72   C   A
+49.49
FFL +45.88 Level 6
+43.16
C D
FFL +42.43 Level 5
FFL +39.38 Level 4
FFL +36.49 Level 3
FFL +33.60 Level 2
FFL +30.71 Level 1
FFL +27.82 Level G

Walpole House   Hallam Court

east elevation

Proposed East Elevation B-B

1:100   0   10m   20m

Elevation 2
立面图 2

COLOR

Buildings are mainly dominated by yellow stone, and in this case, designers use brass as the decoration material for the façade and balcony. Golden architectural appearance and tone blend into the surrounding buildings, and the materials make the building stand out. All these make the building in the corner gleaming.

色彩

本案中建筑主要以土黄色的石材为主，此外，设计师采用黄铜作为立面和阳台装饰材料。金灿灿的建筑外观和色调与周围建筑融为一体，材质上亦使得建筑脱颖而出。所有的搭配使得原本城市角落里的建筑熠熠生辉。

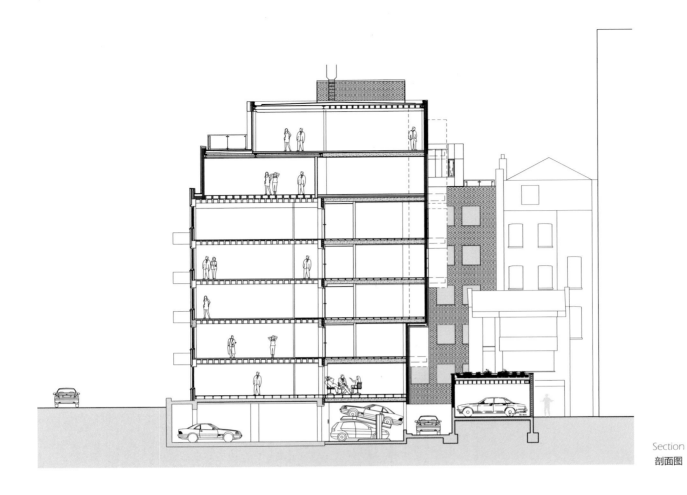

Section

剖面图

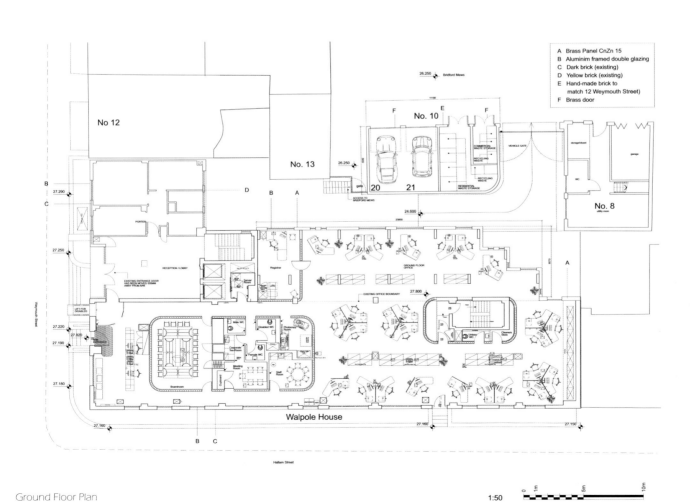

A   Brass Panel CnZn 15
B   Aluminim framed double glazing
C   Dark brick (existing)
D   Yellow brick (existing)
E   Hand-made brick to
      match 12 Weymouth Street)
F   Brass door

No 12

No. 10

No. 13

No. 8

Walpole House

Ground Floor Plan

首层平面图

1:50

SUSTAINABILITY 可持续性

FAÇADE 立面

COLOR 色彩

MATERIAL 材料

One Hyde Park, London, UK
英国伦敦海德公园一号

Architect: Rogers Stirk Harbour + Partners
Client: Project Grande (Guernsey) Ltd
Location: London, UK
Site Area: 7,500 m²
Gross Floor Area: 65,000 m²
Floors: 10, 12, 14, 12
Height: 46 m
Photography: Paul Raftery, Timothy Soar, Edmund Sumner

设计公司：罗杰斯·史达克·哈伯＋合伙人建筑事务所
客户：Project Grande (Guernsey) Ltd
地点：英国伦敦市
占地面积：7 500 平方米
建筑面积：65 000 平方米
层数：10、12、14、12
高度：46 米
摄影师：Paul Raftery、 Timothy Soar、Edmund Sumner

One Hyde Park beside Mandarin Oriental Hotel is a modern residential development near to Knightsbridge, London, on the site of the former Bowater House office building (originally built in the late 1950s). The site is facing south carriage drive in Hyde Park to the north, Knightsbridge to the south, adjacent Wellington Court to the west and Mandarin Oriental Hotel to the east.

文华东方酒店旁的海德公园一号公寓大厦位于伦敦骑士桥附近，此地曾是宝华特公司办公楼的旧址（始建于 20 世纪 50 年代末）。项目北面是海德公园的南车道，南面是骑士桥，西面与惠灵顿法院相邻，东面是文华东方酒店。

The development delivers 86 residential apartments and duplexes and three retail units (at ground-floor fronting onto Knightsbridge) within four interlinked pavilions. These pavilions step up in two-storey increments to give, from west to east, buildings of 10, 12, 14, and 12 storeys. Facilities for residents include a private cinema, a 21m-long swimming pool, a squash court, a state-of-the-art gym, a business suite and meeting rooms. In total, approximately 34,340 m² residential and 836 m² retail space is offered by the development, with 139-vehicle and 114-cycle parking spaces occupying two basement levels.

One Hyde Park seeks to complement the existing streetscape of Knightsbridge and creates a scheme that offers daylight and generous views whilst achieving the necessary degree of privacy for its occupants. The project is designed to be flexible and legible. The residential pavilions are separated by a series of fully-glazed circulation cores, stairs, lifts and lobbies. The passenger cores are used by the development's residents for primary access to the apartments and penthouses, and service cores are used for secondary access by staff and for providing service access to the apartments. The superstructure of the residential accommodation in each pavilion comprises an exposed, pre-cast concrete frame, expressed externally in two-storey elements.

本案位于 4 个相互连接的展馆内，设有 86 套公寓和复式套房、3 间零售商店（位于首层，面对骑士桥）。这些展馆从东到西以两层高度递增，4 栋大楼分别为 10、12、14 和 12 层。公寓为居民提供众多公共设施，包括一家私人电影院、21 米长的游泳池、壁球场、设备先进的健身房、商务套房和会议室。住宅区总面积为 34 340 平方米，零售商店为 836 平方米，占用地下两层的停车场，停车场内带有 139 个停车位和 114 个自行车位。

海德公园一号旨在丰富骑士桥的街景，并为住户提供采光充足、视野广阔和保证隐私的住宅空间。项目的设计灵活而清晰。住宅部分被一系列全玻璃围合的循环核心、楼梯、电梯和大堂分隔。乘客中心主要服务公寓的住户，方便出入公寓和套房；服务中心则为工作人员使用，方便他们为公寓居民提供服务。每部分最上层的公寓由裸露的预制混凝土框架构建而成，在外部表现出两层楼的元素。

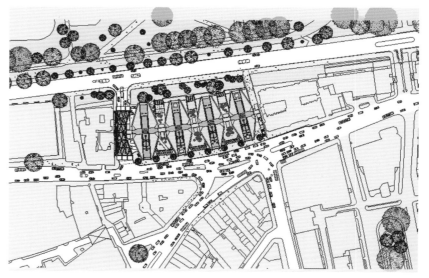

Site Plan
总平面图

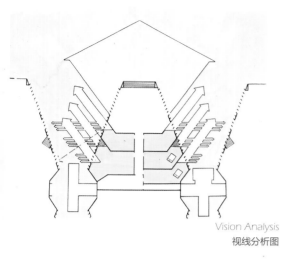

Vision Analysis
视线分析图

Daylight Analysis
日照分析图

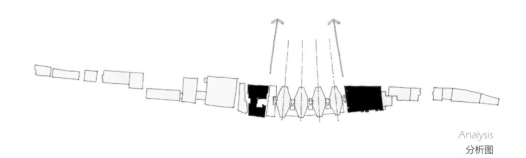

Analysis
分析图

A series of sustainable initiatives have been developed and incorporated into the overall design.

The geothermal strategy uses ground water from an aquifer some 150 m below ground level. Heat added and stored in the aquifer by the building in the summer will be extracted and raised by a heat pump to heat tap water and the building itself in the winter. Balancing heat in and out sustains use of the aquifer for future generations. In the long term, this approach will reduce the primary energy consumption of the building as well as the emission of greenhouse gases and other pollutants. This strategy is to meet the Greater London Authority's minimum requirement to provide a minimum of 10 percent of the development's energy requirements using renewable sources.

Highly-insulated façades will be shaded by using integrated solar blinds and fixed privacy screens. Using solar sensors, the blinds are automatically deployed to reduce solar gains only when required. Daylight is, therefore, optimized whilst spectacular views from rooms are preserved. All perimeter rooms have operable windows for natural ventilation which, when opened, automatically turn off the local cooling.

Green roofs and ground-based planting areas not only provide beautiful surroundings as well as act as a "sponge", greatly reducing the speed of rainwater, with which rainwater runs off the building surfaces before passing into a series of tanks forming a second stage of attenuation and reducing the impact on the existing sewer systems. In addition, rainwater will be collected and stored to be used for irrigation of the planted areas of the scheme.

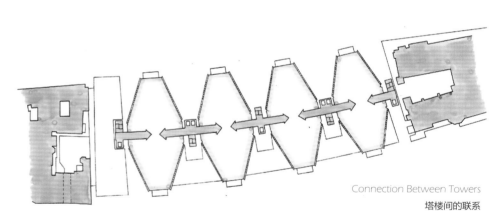

Connection Between Towers
塔楼间的联系

项目的整体设计里还包括了许多可持续发展策略。

地热策略充分利用了地下 150 米的蓄水层的地下水。在夏日，建筑将热量储存在地下蓄水层，冬天则用热泵将其抽提出加热自来水，以及为建筑自身供热。平衡热量的进出，合理而充分，使得蓄水层为后代持续利用。从长远来看，这种方法会减少建筑主要的能源消耗以及温室气体和其他污染物的排放量。这个策略是应大伦敦市政府的能源最低要求而提出的，运用可再生能源以至少符合能源开发要求的 10%。

太阳能百叶窗和固定的私密屏风能为高绝缘立面遮挡阳光。利用太阳能传感器，百叶窗在需要时能自动展开遮挡太阳辐射。因此所有房间都享有充足的日光和广阔的视野。所有外围的房间配置有可操纵的窗口，方便自然通风，打开后，房间里的冷气会自动关闭。

屋顶花园和地表植被不仅具有美化功能，更是一块"海绵"，大大降低了雨水的流速，在雨水流过建筑表面，进入排水管前，屋顶花园和地表植被作为第二阶段的过滤，减轻建筑原有的污水系统的负担。另外，雨水将被收集并储存起来，用于植被的灌溉。

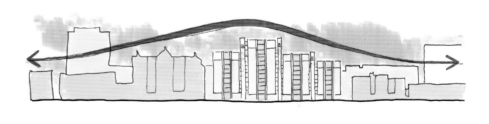

Skyline Diagram
天际线示意图

Pre-patinated copper alloy – the privacy screens were fabricated by using pre-patinated copper alloy. The colouration of the patina was derived from the brickwork infill panels on adjacent buildings. The aspiration for the patina was to capture the rich variegations and textures of the brickwork. The screens are fabricated by using cast stainless steel end brackets, extruded aluminium skeleton, with profiled perforated copper alloy sheets and stainless steel trims.

Steelwork — all visible steelwork is colored in dark grey to match the dark ironwork of the adjacent buildings. The structural system for the circulation cores takes its lateral support via a steel grillage from the adjacent pavilions, with horizontal "Y-shaped" frames which, in turn, offer lateral stability to the staircases, resulting in a two-storey expression of vertical support and horizontal restraint at alternate levels. The nodes and connections are expressed with precision steelwork detailing.

Concrete — all exposed elements of the concrete frame are pre-cast components, cast using bespoke steel shutters. The final finish to the surface is achieved by acid etching which exposes the limestone aggregate and the mica content of the concrete mix. A clear sealer is then applied.

Glass — the majority of the glass used in the development has a low iron content, resulting in a high degree of transparency and avoiding the "green" tint usually experienced in conventional annealed glass.

带有绿锈的铜合金屏风起到保护住户隐私的作用，采用铜绿色是取材于相邻建筑上的砌体填充板。使用光泽材料是希望捕捉到砖墙上丰富的色彩和质感。屏风的材料包括浇铸不锈钢尾架、铝制挤压型骨架、异形穿孔铜合金板和不锈钢装饰。

为了和相邻建筑的黑色钢铁结构相匹配，所有可见的钢结构都漆上了深灰色。流通核心的结构系统以横向支承结构，从钢格床一直延伸到相邻的结构，和"Y"形框架一起提高楼梯的横向支承，创造出两层高的垂直支撑和水平限制。节点和连接点都用精密钢结构打造。

所有暴露在外的混凝土框架都是用预制部件和浇铸定制的钢护窗板建造而成。最后在立面的精制加工上用了玻璃蚀刻法和混合了裸露在外的石灰石和云母的混凝土混合物。最后再加上一层清晰的密封层。

建筑里采用的大部分玻璃都具有低含铁量，这样既能有高透明度，又不用采用常见的退火玻璃的绿色。

FEATURE　特点分析

FAÇADE

The façade system of the residential levels utilizes a series of vertical blade-like elements set within the concrete frame. The two-storey-high screens that form the main façade are of pre-patinated copper alloy, with the patination complementing the coloration of building materials in the immediate area. The primary function of the screens is to provide privacy to residents as well as to control views out of and into the building. These screens also provide solar shading which gives depth and modeling to the façade. The façade system becomes more transparent to views from Hyde Park and Knightsbridge and is predominantly solid when viewed from adjacent properties.

立面

住宅层的立面系统利用一系列垂直的叶片状元素，设置在混凝土框架内。这种两层楼高的带有铜绿的铜合金屏风组成了建筑的主立面，锈色的外观让工作区的建材色彩更为丰富。这些屏风的主要功能是能确保住户的隐私，同时住户可以随时打开屏风欣赏风景。这些屏风具有遮挡日光的作用，它为立面创造了深度与立体感。从海德公园和骑士桥方向看来，建筑的立面变得更为透明，而从相邻建筑看过来却又是实心的。

Elevation 1
立面图 1

Elevation 2
立面图 2

Section
剖面图

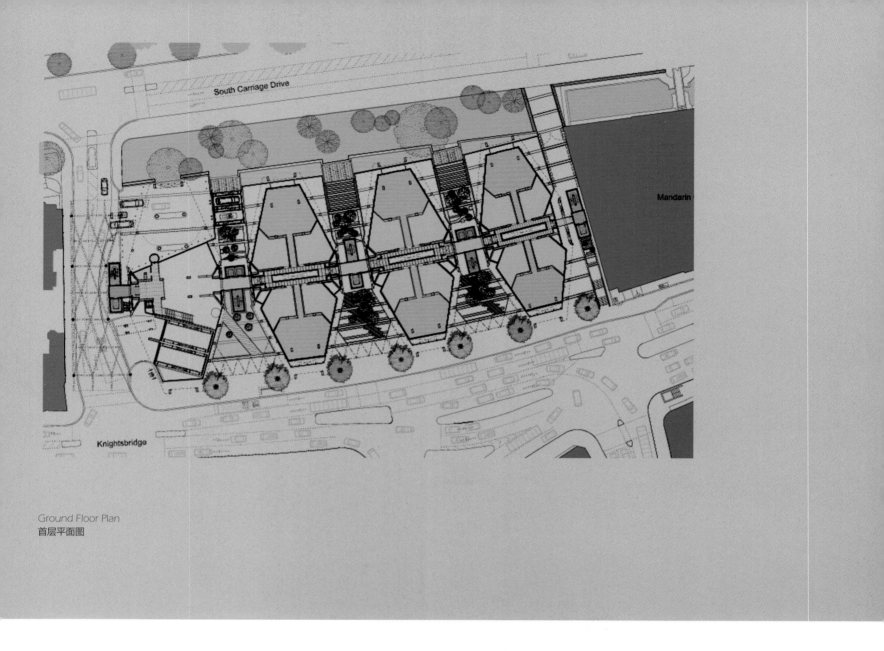

Ground Floor Plan
首层平面图

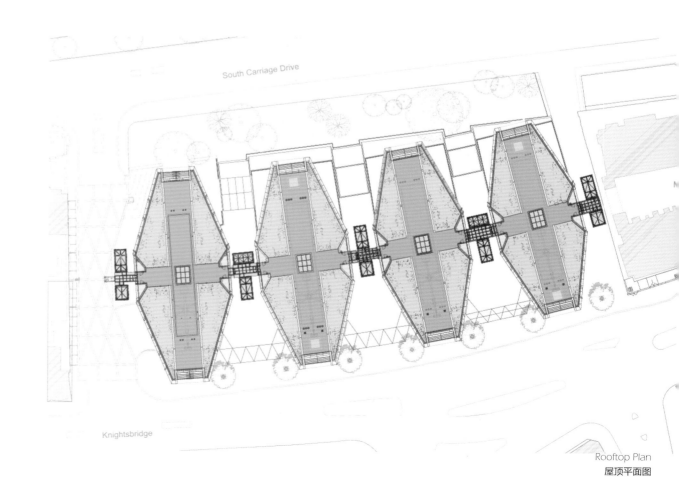

Rooftop Plan
屋顶平面图

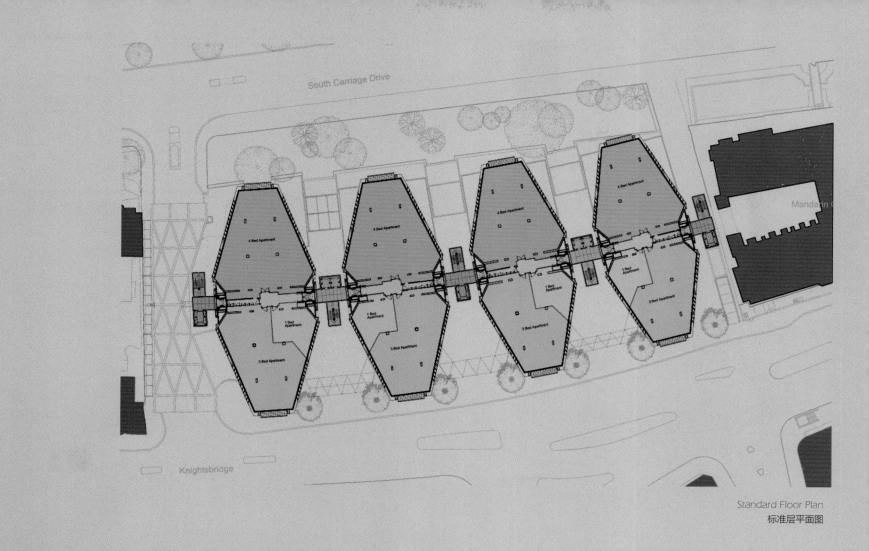

South Carriage Drive

Knightsbridge

Standard Floor Plan
标准层平面图

The top level of each block comprises two-storey penthouse accommodation which relates and responds to the roofscapes of neighboring buildings. The façades of the penthouses have been developed to incorporate the geometry and privacy aspects of the façade system below.

每栋楼的最顶层都有两层楼的复式套房，可欣赏到毗邻建筑的屋顶景观。复式套房的立面系统与下层的具有一致的几何形状和私密性。

FAÇADE 立面

SHAPE 造型

MATERIAL 材料

SUSTAINABILITY 可持续性

High Line 23, New York, USA

美国纽约高线 23 号大厦

Architect: Neil M. Denari Architects
Location: New York, USA
Site Area: 371.61 m²
Gross Floor Area: 4,273.54 m²
Floors: 14
Height: 53.34 m
Photography: Benny Chan / Fotoworks 2012

设计公司：Neil M. Denari Architects
地点：美国纽约市
占地面积：371.61 平方米
建筑面积：4 273.54 平方米
层数：14
高度：53.34 米
摄影：Benny Chan / Fotoworks 2012

High Line 23 is a 14-floor condominium tower that responds to a unique and challenging site directly adjacent to the High Line at 23rd street in New York's West Chelsea Arts District.

本项目是一栋高 14 层的公寓大楼，位于纽约切尔西艺术区 23 号街，紧邻高线公园，这个独特而富有挑战性的地形造就了建筑与众不同的造型。

Partially impacted by a spur from the elevated tracks that make up the High Line superstructure, the site is 40' x 99' at the ground floor. The site and the developer demanded a specific response, yielding a project that is a natural merger between found and given parameters. For the client, the question was how to expand the possible built floor area of a restricted zoning envelope; for the site, a supple geometry must be found to allow a larger building to stand in very close proximity to the elevated park of the High Line. Together, the demands produce a building with one unit per floor and three distinct yet coherent façades, a rarity in Manhattan's block structure.

受到高线公园上层高架轨道结构的影响，首层场地面积只有 40 英尺 x 99 英尺。鉴于场地和开发商的特殊要求，建筑必须在现有给定的参数中寻找突破口，实现双赢。对于客户来说，问题在于如何在场地限制下尽可能扩大建筑面积；对场地来说，必须找到一个灵活的几何结构来使建筑尽可能地接近高线公园的高架结构。综合两者，一栋每层只有一间套房、3 个风格独特连贯立面的建筑就此诞生，并成为了曼哈顿少有的建筑结构。

FEATURE 特点分析

SHAPE

With a custom non-spandrel curtain wall on the south and north façades, and a 3D stainless steel panel façade on the east facing the High Line, the project's geometry is driven by challenges to the zoning envelope on the site and by NMDA's interest in achieving complexity through simple tectonic operations.

造型

建筑的南北立面是定制的非拱肩幕墙，东立面则覆盖了立体不锈钢面板，朝向高线公园。建筑的独特造型正是受到场地限制的驱动，设计师通过简单的构造实现了复杂的结构。

STRUCTURE 结构

VIEW 视野

BALCONY 阳台

FAÇADE 立面

La Taille Vent, Hamburg, Germany

德国汉堡 La Taille Vent 公寓

Architect: Spine Architects
Client: Stilwerk Living GmbH
Location: Hamburg, Germany
Site Area: 3,000 m²
Floors: 8
Photography: Oliver Heissner, Spine Architects

设计公司：Spine Architects
客户：Stilwerk Living GmbH
地点：德国汉堡市
占地面积：3 000 平方米
层数：8
摄影：Oliver Heissner、Spine Architects

Life has come to Hamburg's Hafen City! Although visitors are still confronted by uncountable cranes as well as the building noise that can drown out the sound of the wind, there are clear signs that people are beginning to inhabit in Germany's largest and most exciting building project.

汉堡的港口城即将重新恢复活力！虽然游客还可以看到无数的起重机，建设噪声甚至能掩盖风声，但有明显迹象表明，人们已经开始选择在这个德国最大型、最振奋人心的住宅项目中居住。

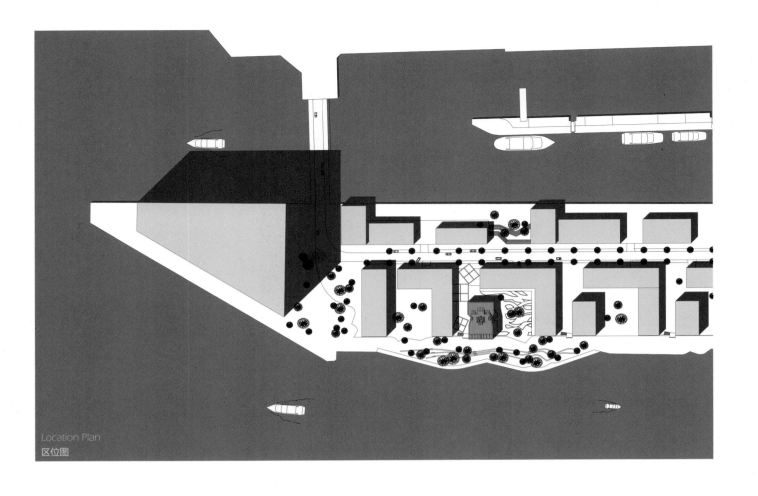

Location Plan
区位图

The development of the "Sandtorkai" in the north forms the prelude to the Hafen City. The shiny new private yachts and more traditional ships are already anchored here and await the good weather.

On the southern Kaiserkai, direct on the banks of the River Elbe is "La Taille Vent". Here every apartment is already inhabited. This architectural structure, with its dynamic style, is one of the most outstanding projects created for this new location.

Here the "sun deck feeling" is guaranteed at every level. The uninterrupted expanse of glass forming the windows opens each flat in its full width to the water, letting in light and air. The architectural structure tapers in the middle, also enabling the flats at the back of the building to enjoy a direct alignment to the river. Each resident has an everlasting, unobstructed view over the River Elbe.

北部的 Sandtorkai 开发项目就是港口城的发展序曲。闪亮的新私人游艇和传统的船舶已经锚定在这里，等待好天气的到来。

La Taille Vent" 项目恰好位于德国汉堡市南部地区 Kaiserkai 的易北河岸，目前已全部入住。建筑结构动感十足，是这个新兴区域里最夺目的项目之一。

公寓的每层楼都享有 "阳光露台" 的感觉。连续延伸的玻璃窗为每户打开广阔的水面视野，使光线和空气进入。建筑结构在中部逐渐变细，如此一来建筑背面的单元也能享有水景视野。每个居民都时时享受易北河的美景。

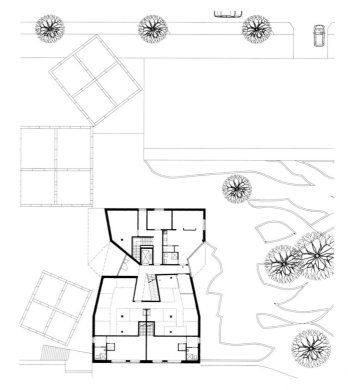

Site Plan
总平面图

Section
剖面图

FEATURE 特点分析

VIEW

"La Taille Vent" means "wind strength" or translated into "formed by the wind" more loosely. The slender structure of this apartment house manifests this by seemingly soaring into the altitude and yet, inclining towards the river. Careful considerations are given to each unit to guarantee a wonderful Harbor scenery.

视线

"La Taille Vent" 意思是"风的力量",或可翻译成"因风而形成"。公寓建筑看似高耸的体量营造出一种修长的结构,并向河边渐渐倾斜。细致的设计确保每个单元都可以尽享港湾美丽的风光。

The other advantage is "waistline" that makes the floor plan become more interesting. The company Spine Architects, attaches great importance to a spacious floor plan. It should be a reflection of the modern, open, harmonized style of living. Here the living, cooking and dining areas are together and orientate towards the River Elbe. The rooms can be separated off or opened out to each other. The bedrooms and bathrooms with their more private characters are situated on the northern side of the building.

Every apartment possesses a large balcony or terrace overlooking the River Elbe, accessible through sliding doors.

The outcome of the Hafen City for Hamburg is a completely new urban center providing work and accommodation. Adjacent to "La Taille Vent" will be the Elbphilharmonie, Hamburg's new cultural flagship.

项目的另一个优势是体量的"腰线",它形成了有趣的楼层布局。设计公司 Spine Archtects 非常重视宽敞的楼层布局,认为必须和现代、开放、和谐的生活方式相适应。于是,起居室、厨房和餐厅均面向易北河。房间可以彼此分离或合并,卧室和浴室则相对隐蔽,被设置在建筑的北面。

每个公寓都拥有一个大阳台或露台俯瞰易北河,并可以通过滑动门自由出入。

如今,港口城已经成为汉堡全新的城市中心,提供各种工作和住宿机会。"La Taille Vent"公寓附近将建成汉堡的新文化旗舰中心——Elbphilharmonie。

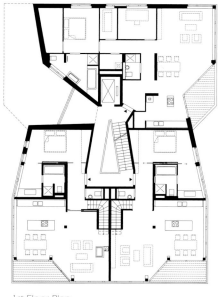

1st Floor Plan
1 层平面图

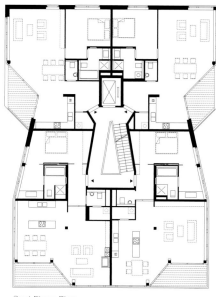

2nd Floor Plan
2 层平面图

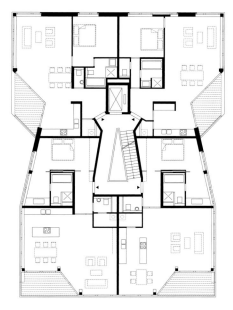

3rd Floor Plan
3 层平面图

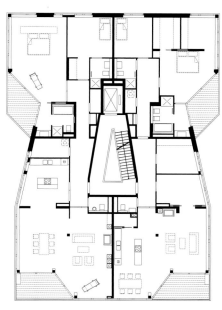

4th Floor Plan
4 层平面图

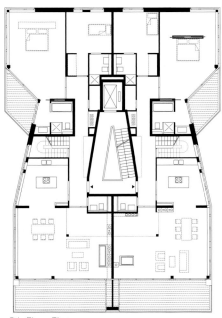

8th Floor Plan
8 层平面图

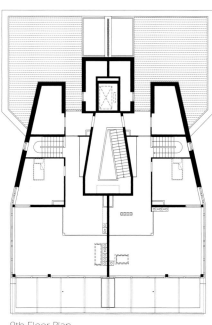

9th Floor Plan
9 层平面图

MATERIAL 材料

BALCONY 阳台

OPEN SPACE 开发空间

SUSTAINABILITY 可持续性

Ex Rossi Catelli Area Residential Building, Italy

意大利 Ex Rossi Catelli 区公寓

Architect: Cino Zucchi Architetti
Client: Parmavera
Location: Parma, Italy
Site Area: 12,256 m²
Floors: 5
Height: 19.35 m
Photography: Cino Zucchi

设计公司：Cino Zucchi Architetti
客户：Parmavera
地点：意大利帕尔马市
占地面积：12 256 平方米
层数：5
高度：19.35 米
摄影：Cino Zucchi

The site of the project is situated in the northwest corner of the new plan, straddled between the bustle of daily life of the avenues that connect the area to the center and the quiet park. This invites and gives synthetic shape to the needs of contemporary living in a place of high environmental quality.

本项目位于新规划场地的西北角，连接着日常生活的喧嚣场所、中心区和宁静的小公园。这些环境条件造就了建筑的组合造型，以高环境质量的空间来满足当代生活的需求。

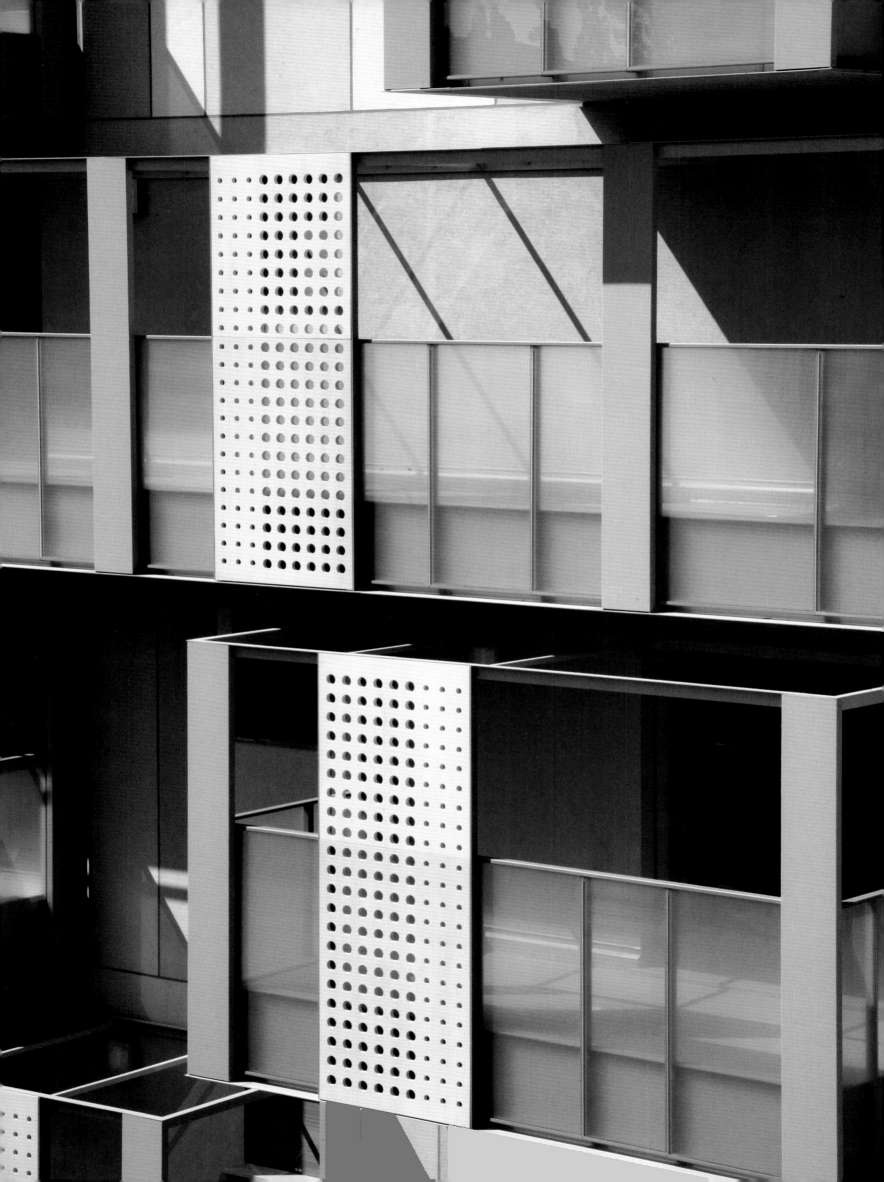

BALCONY

To bring ventilation and lighting to interior, each apartment is configured with a large balcony with green glass railing, which severs to filter sunlight as a cozy outdoor space. In addition, a porous panel is set on each balcony as high as the ceiling that helps block view from neighborhood without affacting functions of the balcony.

阳台

为给封闭的室内带来通风和采光，每户都配置了大阳台，以绿色玻璃作为栏杆，既可为室内过滤阳光，又能给用户带来一处舒适的户外空间。另外，阳台上设置了等同天花板高度的多孔面板，不仅不影响阳台的通风和采光，而且确保了室内的私密性。

The open courtyard setting of the buildings protects a central green area which the large terraces face. The positionings of these buildings have been carefully studied in relation to the movement of the sun so as to create "open air living rooms" which allow the lodgings to be lived in all seasons of the year. All communal entrances can be accessed directly from the perimeter of the site and are connected to the vast area of the building's garden.

建筑的开放庭院保护了阳台面向的中央绿色区域。这些建筑的位置根据太阳的运动规律而布置，以创造"露天客厅"，让住户在一年四季都能舒适地生活。从场地的外围通过公共入口就能进入建筑，这些入口同建筑的大面积花园相连。

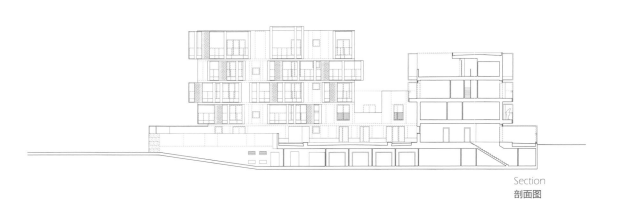

Section
剖面图

East Elevation
东立面图

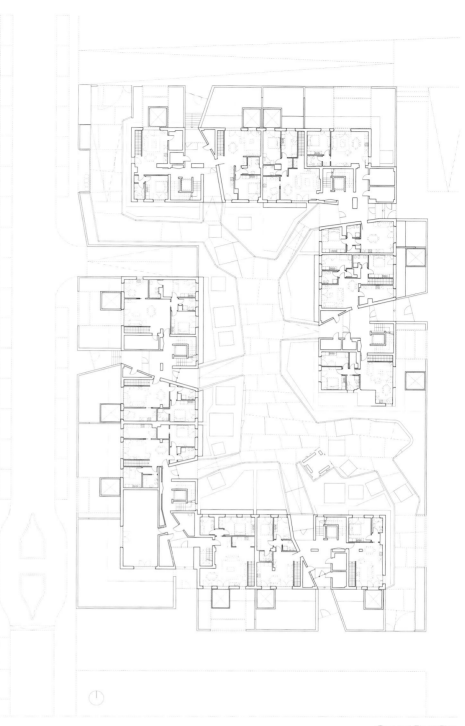

Ground Floor Plan
首层平面图

2nd Floor Plan
2 层平面图

FAÇADE 立面

SHAPE 造型

LAYOUT 布局

SUSTAINABILITY 可持续性

Building 29 Living Units and Shops, Hérault, France

法国埃罗 29 房公寓和商店大楼

Architect: MDR Architectes
Client: Corim Promotion
Location: Montpellier, Hérault, France
Site Area: 2,756 m²
Floors: 5
Photography: Stéphane Chalmeau
Realization: 2012

设计公司：MDR Architectes
客户：Corim Promotion
地点：法国埃罗省蒙彼利埃市
占地面积：2 756 平方米
层数：5
摄影：Stéphane Chalmeau
完成时间：2012 年

This project is located on the outside rim of the new Port Marianne neighborhood, and the quality of the location is remarkable in many ways: in front of the avenue Raymon Dugrand. This location makes a flagship project for this new neighborhood with a high-quality urban environment; on the northern front, the land is bordered by a beautiful alley of century-old trees and on the eastern front the new landscaped park offers a one-of-a-kind view.

本案位于新玛丽安港口区的边缘位置，这个位置具有明显的优势：面对着 Raymon Dugrand 林荫大道，为本案的旗舰项目所在的街区提供了高质的城市环境；北部的区域以一条植有百年树龄树木的小巷作为边界，东部的区域有一座新建的园林公园，为公寓提供独一无二的景观。

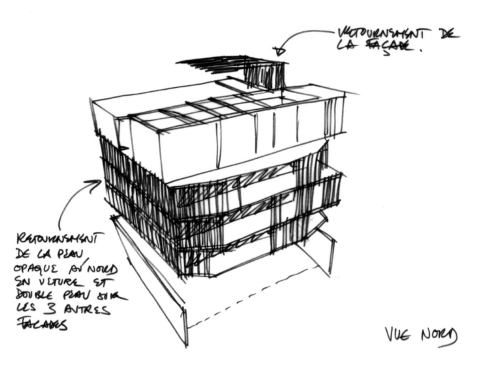

RETOURNEMENT DE LA FACADE.

RETOURNEMENT DE LA PEAU OPAQUE AU NORD EN TOITURE ET DOUBLE PEAU SUR LES 3 AUTRES FACADES

VUE NORD

Sketch 1
草图 1

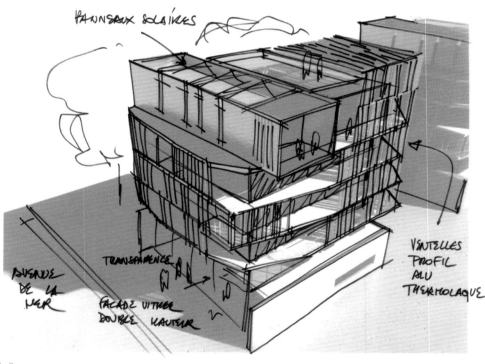

PANNEAUX SOLAIRES

AVENUE DE LA MER

TRANSPARENCE

FACADE VITREE DOUBLE HAUTEUR

VENTELLES PROFIL ALU THERMOLAQUE

Sketch 2
草图 2

FEATURE 特点分析

SHAPE

The building appears to be two separated parts. The upper one features in irregular floors edged in white tile with large grey windows sandwiched between two floors. The white tile and dark glass form a strong contrast. There are groups of linear elements connecting each floor randomly, which screen direct views and bring a kind of freshness to the building. The lower part of the ground floor shows an opposite image against upper lightness. The thick brick wall structure gives a stable and solid impression. The two parts are connected by glass axis seamlessly, though looked separately.

造型

建筑在视觉上分为两部分，上部的不规则楼层由白色瓷砖镶边，与夹在中间的深色玻璃窗形成鲜明对比。各组线性元素以随机的位置联系各个楼层，起到一定屏蔽作用的同时，也为建筑整体带来别样的新鲜感。下部底层则和上层的轻盈感截然相反，厚实的砖墙结构给人稳重、坚固的印象。上下层由玻璃中轴无缝相接，一分为二又浑然一体。

The first building has 13 living units from the first to the fifth floor, going from 1-bedroom to 4-bedroom. The ground floor is devoted to the building entrance and business openings on the Dugran Avenue and the central esplanade between the buildings. The first level is divided between apartments and shops' upper levels.

The second building hosts 16 apartments from 1-bedroom to 3-bedroom from the ground floor up to the fifth floor.

Ground floor apartments are raised-up from the pedestrian areas in order to preserve the residents' privacy. It is however accessible by the elevator.

The rest of the ground floor has shops on its western half facing the central courtyard. A ramp accessing the underground parking is located under the raised embankment of the ground floor apartments.

第一栋建筑第二层至第六层共有 13 间住房，从一居室到四居室不等。首层除了用作建筑入口，同时也包括 Dugran 大道上的商业门面以及两栋建筑之间的空地。二层被划分为住宅和商店的上层。

第二栋建筑同样是 5 层共有 16 间住房，从一居室到三居室不等。

位于首层的公寓高于人行道，以保护住户隐私，另外也配有电梯。

首层西半部设有商店，面向中央庭院。首层公寓高起路堤下方设有通往地下停车场的坡道。

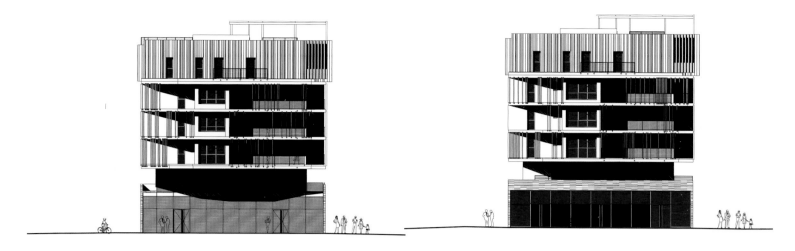

West Elevation
西立面图

South Elevation
南立面图

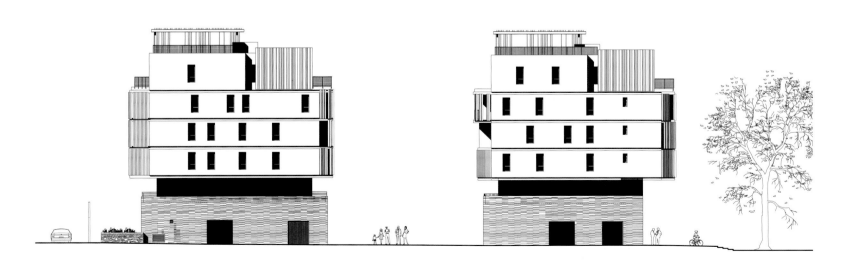

North Elevation
北立面图

The conception of the building's attics is voluntarily innovative and differs from previous projects: made of two boxes placed freely on the bodies of the building anchoring plots, the attic aspires to more freedom. These volumes also contrast by their coverings: fitted with fine vertical slates, the attic becomes a changing object, where brown tones dominate, giving it a lighter feel when it is in fact a solid construction.

建筑的阁楼源自设计师的自主创新，并且不同于以往的项目：两个箱子式的阁楼自由地放置在建筑的主体上，希望能更加突显自由的形态。阁楼的体量也因不同的覆层而形成对比：安装上大小适中的垂直板条，阁楼外观立即呈现出多样化的效果；以棕色为主的色调呈现出更轻盈的感觉，而其实际上是个很坚固的结构。

Section 1
剖面图 1

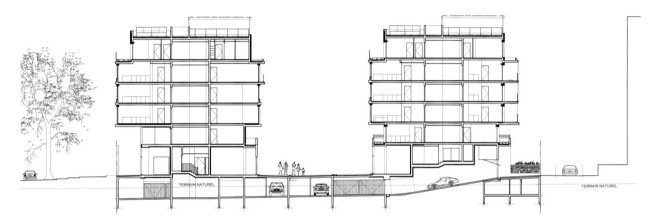

TERRAIN NATUREL

TERRAIN NATUREL

Section 2
剖面图 2

Section 3
剖面图 3

Delimited by the floorboards of the higher levels Ground + 2 to Ground + 5, the buildings' main structure is detached from the base by the height of the first floor (ground +1) looking like a hollow joint and lighting up the volume of the buildings.

To contrast with the full mass of the base, we chose to underline the horizontal layers of the various floors. These bordering tiles undulate in different directions from one floor to the next; these folds create a vibration and a richness of changing shadows that give life to the main structure of the building and break the systematic effect. The apartments' façades rise up in the background with the loggias. This impression of deepness is strengthened by the colors where the white tiles stand out on the darker colors of the façades where windows fit in perfectly. The loggias protect the living units from the glaring sun and offer a nice outside living area on the outside. To reinforce the protection of these loggias from the sun and neighbors' view, randomly placed vertical slates modulate the filled-out and hollow spaces. These slates are thrown from tiles to tiles in areas where they are parallel; established as destructuring elements on the façades and particularly the north one that remains quite solid and protective due to its orientation, these mobile slates connect the main structure of the building. Thus, the base remains solid while offering a more ethereal material, playing with shadows and lights depending on the time of day.

The northern façade expresses this cohesion with a more closed-off and protecting cover giving an impression of neatness overlooking the northern square.

建筑主体由地基首层（1层）和高层的楼面（2层到5层）划分为两部分，看起来像一栋连接的中空体量，这也是本案的一大特色。

为了和首层的结实结构形成对比，设计师选择强调各楼层的水平层。不规则的镶边瓷砖从不同的方向连接各个楼层，这些褶皱产生的动感和丰富的阴影变化为整座建筑带来了活力，打破了系统效应。建筑的背部立面设置了凉廊，白色和深色瓷砖间隔出现加深了建筑立面的深度感，和窗户组成和谐的整体。凉廊能保护公寓免受刺眼阳光的照射，并提供宜人的外部生活区。为了加强凉廊阻挡阳光和隔壁的视线，随机放置的垂直板条能起到调节的作用，或补充或留出空间。这些板条平行连接到每层楼的部分瓷砖，作为立面上可移动的分散元素连接起整个建筑的主体。特别是北立面，因为其朝向需要加强其稳固性和保护性，因此，建筑地基依然保留稳固的特性，使用一种更轻质的材料，能在一天的不同时段产生不一样的光影效果。

北立面表达出一种凝聚力和整洁感，就像是一个封闭的保护层，俯视着北广场。

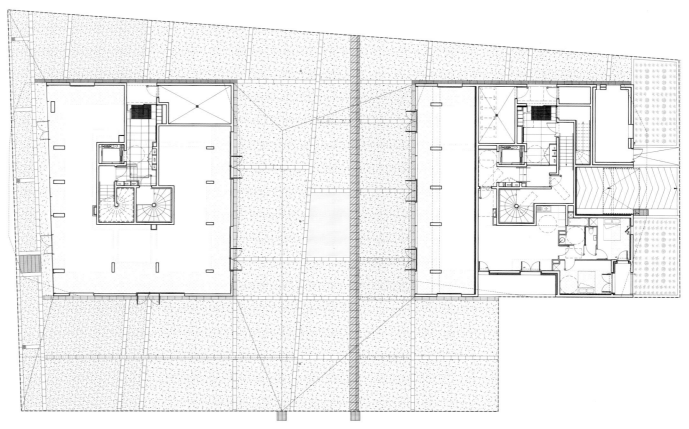

Ground Floor Plan
首层平面图

The pergola roofs are designed as an extension of the vertical slates that cover the southern façades of the supporting bases. Implanted as large triangular shapes, truly shielding the units from the sun, the random organization of the horizontal vents is similar to the vertical portions; these pergolas can discreetly absorb the technical units, necessary for solar panels for heating water and the elevator shaft.

They offer a beautiful protection from the sun for the large accessible patios for the living units on the 5th floor. They play a unifying role for the building bases on the East/West axis along the park; these pergolas are ready to welcome climbing vegetation.

凉亭屋顶被设计成垂直板条的延伸，覆盖到南立面的衬底。屋顶被设计成大三角造型，能真正屏蔽照入公寓内的阳光。随机设置的水平通风口和垂直部分相类似。这些凉亭可以为太阳能电池板吸收能量，以满足热水和电梯井的需求。

屋顶为住在五楼的居民提供了一处能免受阳光辐射的美丽花园，并与沿着公园走向的建筑底层的西/东轴联系在一起。凉亭已经准备迎接攀援植物的生长。

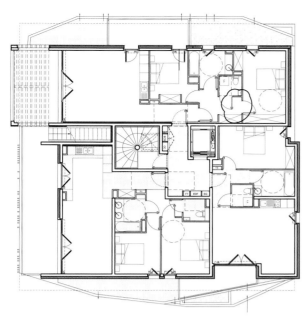

Floor Plan 4
楼层平面图 4

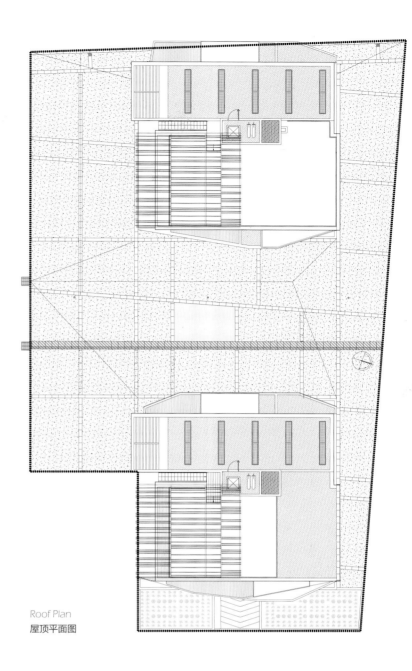

Roof Plan
屋顶平面图

Floor Plan 3
楼层平面图 3

Floor Plan 1
楼层平面图 1

Floor Plan 2
楼层平面图 2

LAYOUT 布局

FAÇADE 立面

SUSTAINABILITY 可持续

MATERIAL 材料

Basket Apartments in Paris, France

法国巴黎篮子学生公寓

Architect: OFIS
Client: Regie Immobiliere de la Ville de Paris
Location: Paris, France
Site Area: 1,981 m²
Gross Floor Area: 8,500 m²
Floors: 10
Height: 29.2 m
Photography: Tomaz Gregoric

设计公司：OFIS
客户：Regie Immobiliere de la Ville de Paris
地点：法国巴黎市
占地面积：1 981 平方米
建筑面积：8 500 平方米
层数：10
高度：29.2 米
摄影：Tomaz Gregoric

The parcel has a very particular configuration: 11m in width and extending approximately 200m north-south. This foreshadows the importance of processing the eastern façade overlooking the extension of the street Des Petits Ponts which hosts the tram and both cyclist and pedestrian walkways.

项目所在的场地非常特殊：宽 11 米，向南北方向延伸约 200 米。这决定了东立面的重要性：因为从这里可以俯视 Des Petits Ponts 街，电车道、自行道和人行道都穿过这条街。

Location Plan
区位图

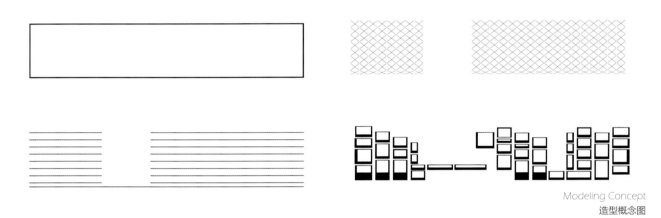

Modeling Concept
造型概念图

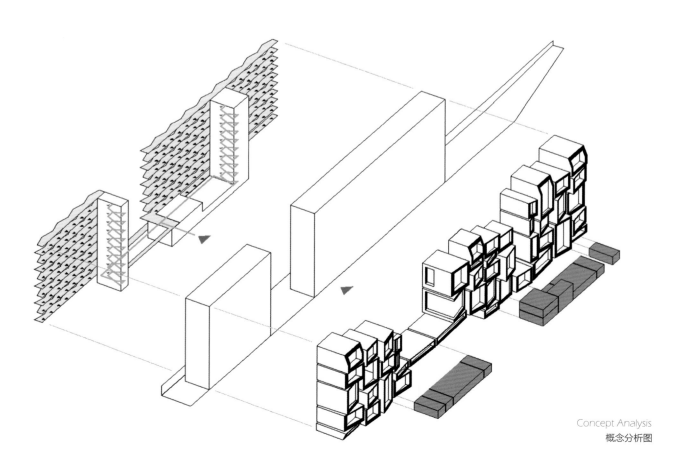

Concept Analysis
概念分析图

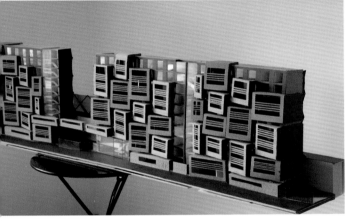

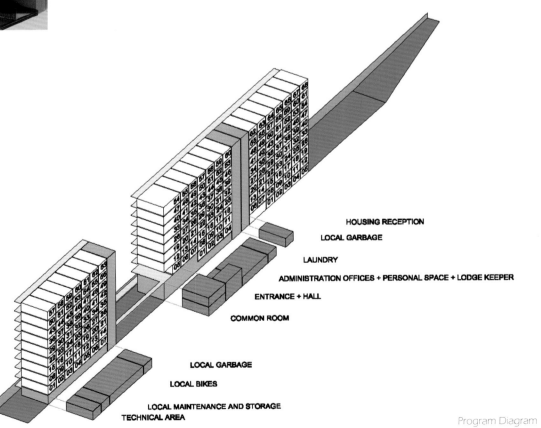

HOUSING RECEPTION
LOCAL GARBAGE
LAUNDRY
ADMINISTRATION OFFICES + PERSONAL SPACE + LODGE KEEPER
ENTRANCE + HALL
COMMON ROOM

LOCAL GARBAGE
LOCAL BIKES
LOCAL MAINTENANCE AND STORAGE
TECHNICAL AREA

Program Diagram
项目分析图

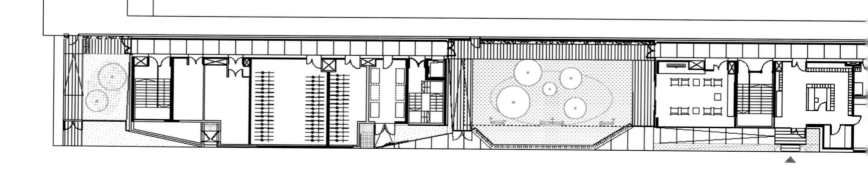

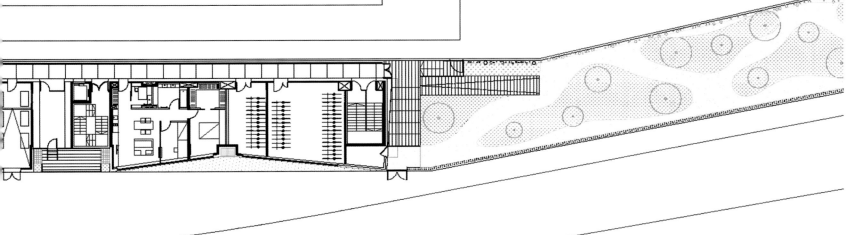

Site Plan
总平面图

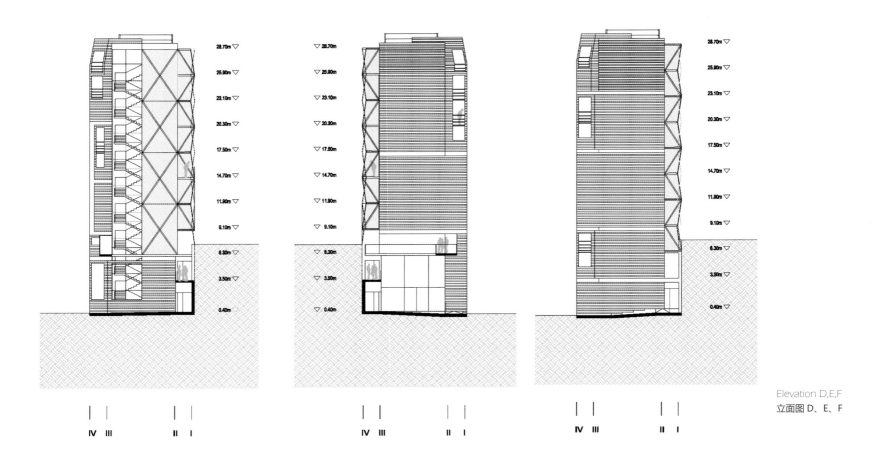

Elevation D,E,F
立面图 D、E、F

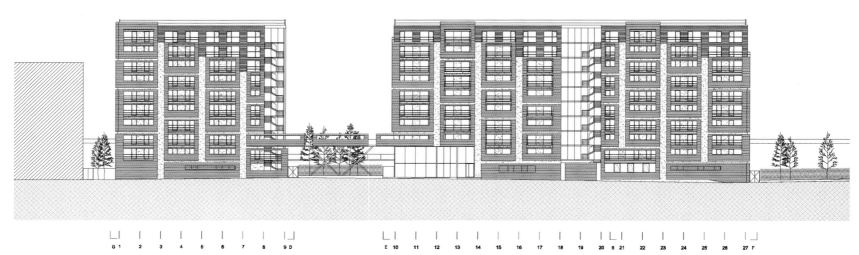

Elevation H-H
立面图 H-H

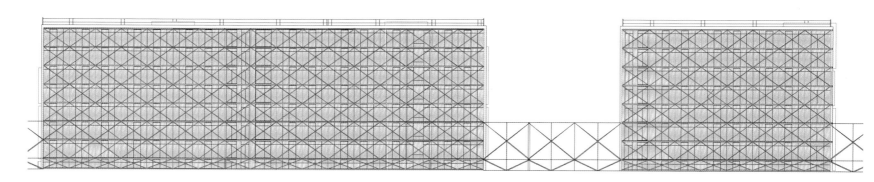

Elevation I
立面图 I

The long volume of the building is divided into two parts connected with a narrow bridge. Between two volumes there is a garden. The building has 11 floors: a technical space in the basement, shared programs in the ground floor, and student apartments in the upper nine floors. The layout is very rational and modular.

The major objective of the project was to provide students with a healthy environment for studying, learning and meeting. Along the football field is an open corridor that overlooks the field and creates a view to the city and the Eiffel Tower. This gallery is an access to the apartments providing students with a common place. All the studios are the same size and contain the same elements to optimize design and construction: an entrance, bathroom, wardrobe, kitchenette, working space and a bed. Each apartment has a balcony overlooking the street.

建筑的长型体量被分为两部分，以一座窄桥相连。两栋体量之间是一个花园。建筑共 11 层，包括一层地下设备室，首层是公共空间，上面的 9 层则是学生公寓。整体布局合理且模块化。

本项目的主要目的是为学生提供健康的研究、学习和交流场所。一条开放式的走廊沿着足球场而建，能俯瞰整个球场，并可远眺城市美景和埃菲尔铁搭。走廊同时也是进入公寓的必经之路，为学生们提供一处公共空间，所有的单元大小相同，内部元素相同，其结构是设计与施工的最完美结合，包括：入口、卫生间、衣柜、厨房、工作空间和一张床。每间公寓都设有一个阳台，提供绝美街景。

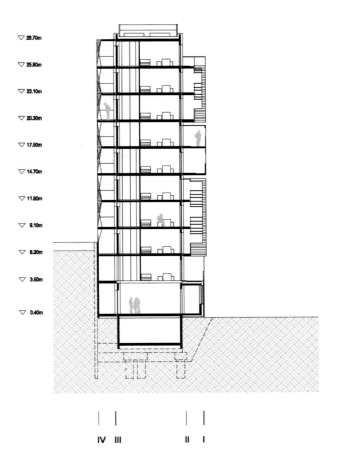

Section B-B
剖面图 B-B

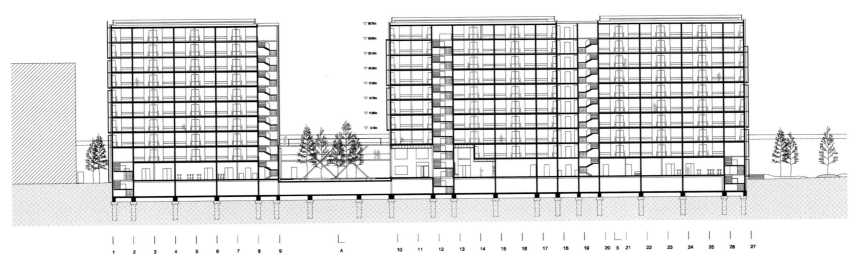

Section C
剖面图 C

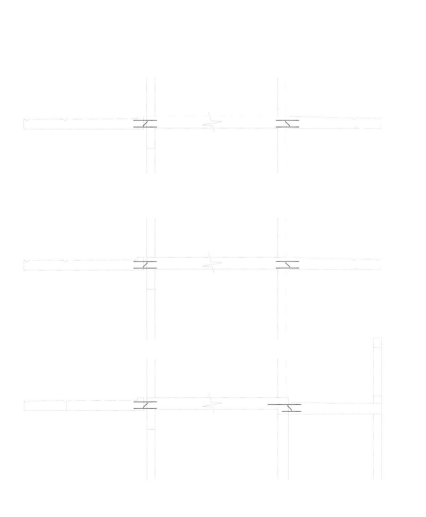

Detail 1
细节图 1

FEATURE 特点分析

FAÇADE

Narrow length of the plot with 10 floors gives to site a significant presence. Each volume contains two different faces according to the function and program.

The elevation towards the street Des Petits Ponts contains studio balconies-baskets of different sizes made from HPL timber stripes. They are randomly oriented to diversify the views and rhythm of the façade. Shifted baskets create a dynamic surface while also breaking down the scale and proportion of the building.

The elevation towards the football field has an open passage walkway with studio entrances enclosed with a 3D metal mesh.

立面

狭长的场地和 10 层的高度是建筑的关键因素。每栋体量拥有两种不同的表面，根据功能和规划进行布局。

朝向 Des Petits Ponts 街的立面是带有篮子状阳台的工作室，尺寸各不相同，立面是由 HPL 条纹木材组成。各单元的结构采用随机朝向，让各单元的视野和建筑的立面都呈现出多样化特征。各单元都如同一个个随意摆放的"篮子"，为立面带来活力，也打破了建筑固有的规模和比例。

朝向足球场的立面有一条开放的走廊，和工作室的阳台一样被立体金属网所围合。

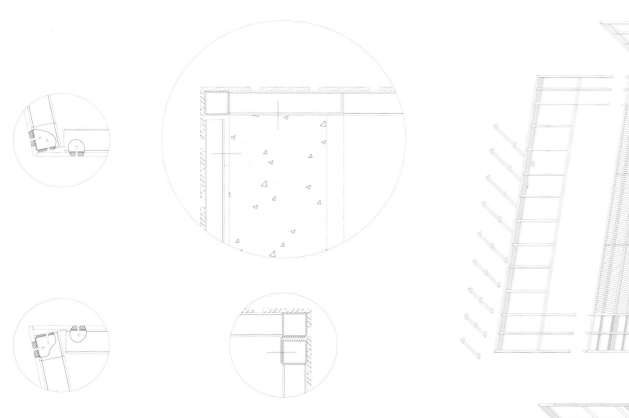

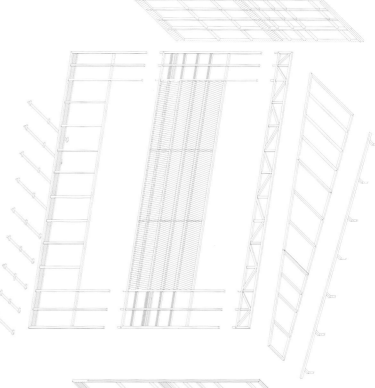

Detail
细节图

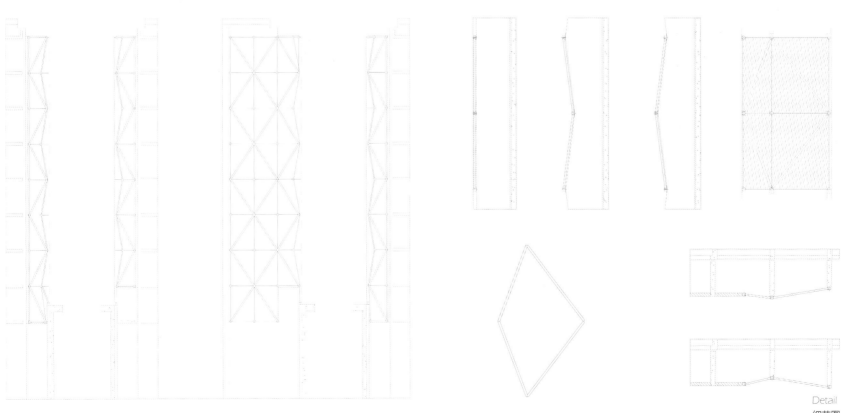

Detail
细节图

Accommodations are cross ventilating and allow abundant day lighting throughout the apartment. External corridors and glass staircases also promote natural lighting in the common circulation, affording energy while also creating comfortable and well-lit social spaces. The building is insulated from the outside with an insulation thickness of 20 cm. Thermal bridge breakers are used on corridor floors and balconies to avoid thermal bridges. Ventilation is controlled by double flow mechanical ventilation, providing clean air in every apartment with an optimum temperature throughout the year. The incoming air also reuses heat from the exhaust air. The roof is covered with 300 m² of photovoltaic panels to generate electricity. Rainwater is harvested on site in a basin pool used for watering outdoor green spaces.

整栋公寓对流通风，并拥有丰富的自然光照。外部走廊和玻璃楼梯也可以起到促进光照循环的作用，同时提供能量，创造一个舒适又明亮的社交空间。建筑通过厚达 20 厘米的绝缘板与外界隔绝。热桥断路阀用于走廊的地板和阳台，以避免热桥。通风由双流机械通风控制，并以最佳的温度为公寓提供清新的空气。传入的空气从废气中重新利用热能。屋顶上覆盖着 300 平方米的太阳能光伏电池板以发电。建筑同时利用水池来收集雨水，用于户外绿色空间的浇灌。

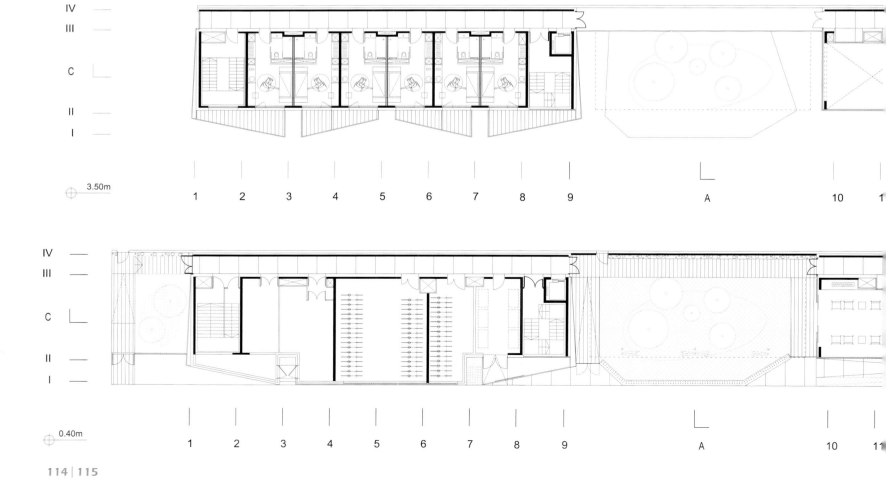

1st Floor Plan
1层平面图

13   14   15   16   17   18   19   20 B 21   22   23   24   25   26   27

Ground Floor Plan
首层平面图

13   14   15   16   17   18   19   20 B 21   22   23   24   25   26   27

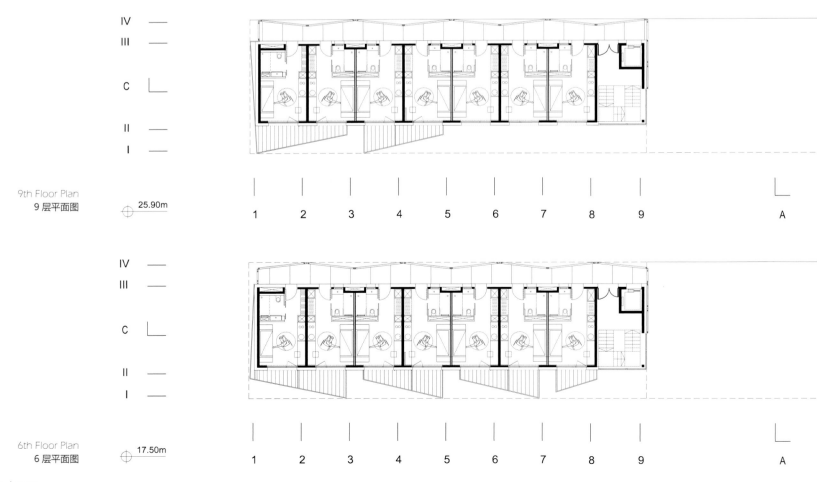

9th Floor Plan
9 层平面图   ⊕ 25.90m

1   2   3   4   5   6   7   8   9   A

6th Floor Plan
6 层平面图   ⊕ 17.50m

1   2   3   4   5   6   7   8   9   A

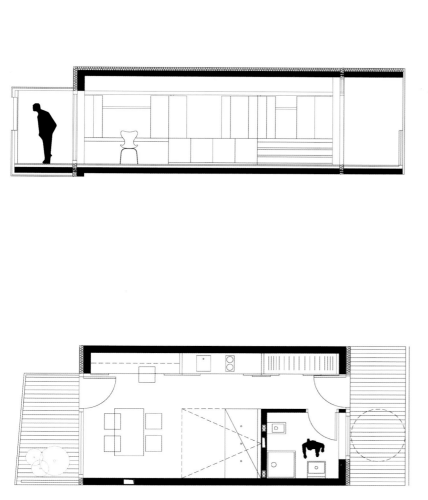

Apartment Units Plan
公寓户型图

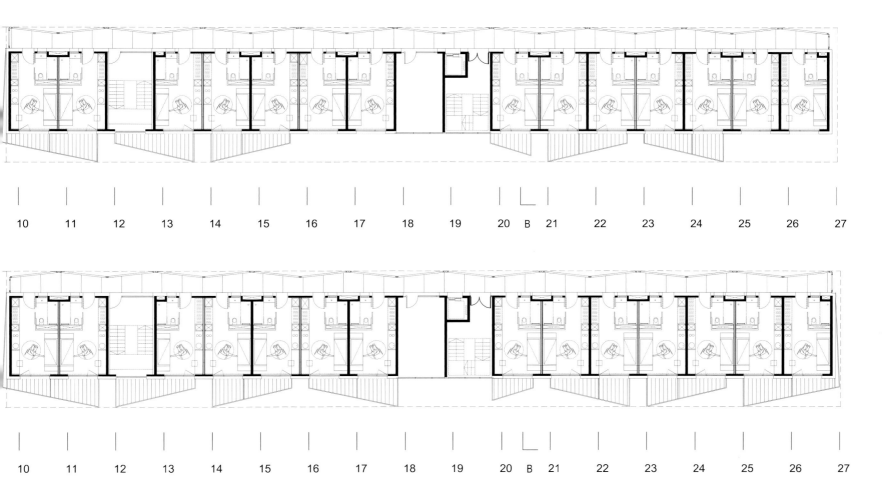

| 10 | 11 | 12 | 13 | 14 | 15 | 16 | 17 | 18 | 19 | 20 | B | 21 | 22 | 23 | 24 | 25 | 26 | 27 |

| 10 | 11 | 12 | 13 | 14 | 15 | 16 | 17 | 18 | 19 | 20 | B | 21 | 22 | 23 | 24 | 25 | 26 | 27 |

FAÇEDE 立面
COLOR 色彩
ECOLOGY 生态
MATERIAL 材料

Ginkgo Project

银杏公寓

Architect: Casanova+Hernandez Architects
Location: Beekbergen, the Netherlands
Site Area: 2,000 m²
Statue: Completed
Photography: Christian Richters, John Lewis Marshall, Casanova + Hernandez Architects

设计公司：Casanova+Hernandez Architects
地点：荷兰贝克贝亨市
占地面积：2 000 平方米
状态：已建成
摄影：Christian Richters、John Lewis Marshall、Casanova + Hernandez Architects

The project is located near the natural park of Veluwe in the Netherlands with views over an old church and the central park of the small town.

"Ginkgo" project explores the possibilities of providing affordable housing for different target groups by its compact housing complex, which is physically and visually integrated in its context.

项目位于荷兰 Veluwe 自然公园附近，能看到旧教堂和小镇的中心公园。

银杏公寓借紧凑的住宅综合体探索了为不同人群提供经济适用房的可行性。住宅在空间上视觉上都融入了周围环境。

COLOR

The printed glazed façade works as a virtual green façade that integrates the building into the greenery of the park and reduces its visual impact in the surroundings, thus giving to the building an iconic image of lightness and immateriality.

色彩

喷涂光滑立面如同虚拟绿色立面，使建筑融入公园的绿化中，减少了它对周边的视觉压迫。也赋予建筑以轻盈和非物质的标志形象。

Site Plan
总平面图

The project is based on a careful dialog between two skins that respond to the border conditions on both sides of the location.

The "Green Skin"

The transparency of the façade facing the park including the long balconies along the apartment block and generous terracing of the park-houses connects the dwellings both visually and physically with the park.

The glazed façade has been specially designed with a print of Ginkgo Biloba tree leaves of different green and yellow tones that react to the constant changing light of the sky creating very special effects, reflections, shadows and silhouettes, depending on the time of the day and the season of the year.

This enables privacy along the balconies and terraces, while ensuring visual connection between the outdoor spaces and the park.

Almost each printed panel of the façade is unique in order to avoid visual repetition, creating a natural and organic continuous image of vegetation that wraps across the whole façade.

The "Urban Skin"

This skin wrapping the building along its front responds to the urban character of its context. The proportions of the openings of the façade and the use of brick that create a more massive appearance to the building help the project to establish a subtle dialog with the existing post-war buildings in the neighborhood.

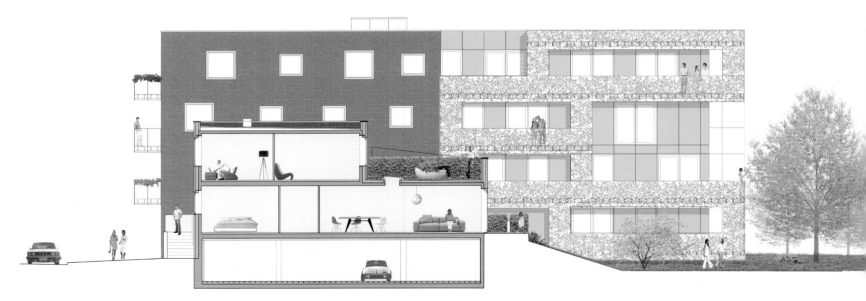

Section
剖面图

项目与场地两侧的边界条件相适应，在两种表皮之间形成对话。

绿色表皮

面向公园的通透立面有着附在公寓的长长阳台。公园和房子间的宽敞走道从视觉上外层和物理上联系着公园和住宅。

玻璃里面上印有不同黄、绿色调的银杏树叶图案，与天空的光线相呼应，根据不同的季节和时间形成了不同的效果。

几乎每扇喷涂立面都是独特的，以避免视觉重复，创造了自然有机的植物图像，包裹整个立面。

喷涂光滑立面如同虚拟绿色立面，使建筑融入公园的绿化中，减少了它对周边的视觉压迫。也赋予建筑轻盈和非物质的标志形象。

"城市肌肤"

包裹建筑前部的表层回应着文脉中的城市要素。立面的开窗比例和砖块的运用形成更巨大的建筑外形使得建筑与毗邻的战后遗留建筑建立微妙的对话。

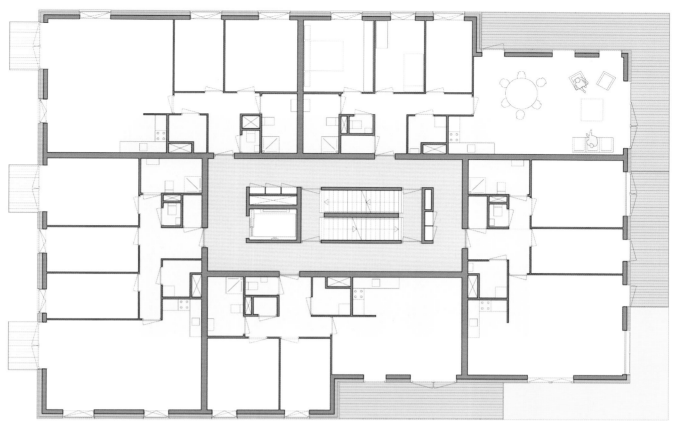

The remarkable high density of the project compared with the neighbor detached and semi-detached houses, not only creates high quality affordable apartments for young people in the town, but also reduces in more than 80% the footprint of the building, minimizes its environmental impact and preserves the nearby valuable natural environment.

The project is designed as a very compact building with a minimum façade surface and a maximum thermal isolation value to minimize the thermal loses in winter, while it is provided with sun screens and long balconies along the south and west façades that work as natural solar protection in summer.

项目相较于周边独立或半独立住宅的惊人的高密度不仅为小镇的年轻人提供了高品质的经济适用房，还减少了 80% 以上的占地面积，使其对环境的影响最小化，保护了附近宝贵的自然环境。

项目设计非常紧凑，将立面最小化以取得热绝缘的最大值，使冬季的热量损失降到最小。同时建筑的南侧和西侧设计有遮阳百叶和长阳台，在夏季提供了自然的遮阳防护。

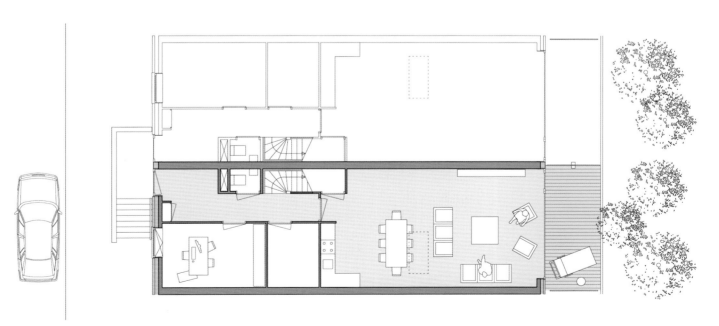

FAÇADE 立面

COLOR 色彩

SUTAINABILITY 可持续性

SHAPE 造型

Darsena Lot 4 Apartment Building on the Harbor, Ravenna , Italy

意大利拉文纳 Darsena 4 号场地港区公寓大楼

Architect: Zucchi & Partners (Cino Zucchi, Nicola Bianchi, Andrea Viganò)
Client: Iter COOPERATIVA RAVENNATE INTERVENTI SUL TERRITORIO
Location: Ravenna , Italy
Site Area: 7,434 m²
Floors: 11
Height: 40.4 m
Photography: Cino Zucchi

设计公司 : Zucchi & Partners (Cino Zucchi, Nicola Bianchi, Andrea Viganò)
客户 : Iter COOPERATIVA RAVENNATE INTERVENTI SUL TERRITORIO
地点 : 意大利拉文纳市
占地面积 : 7 434 平方米
层数 : 11
高度 : 40.4 米
摄影 : Cino Zucchi

The new residential building is part of a large urban renewal project next to the Ravenna Station on both sides of an artificial canal serving as a harbor for the industries of the area. The overall plan by Boeri Studio envisages a new park parallel to the water and a series of rather tall volumes along the waterfront, which currently belongs to the harbor precinct but should in time become open to the public. Within the uncertainty of this "Sliding Doors" future, we designed a double-faced residential complex, relating to the existing city fabric but ready to open toward the water edge and its possible future transformation into a promenade.

本案的新住宅楼是市区重建项目的一部分，它毗邻拉文纳火车站，旁边的人工运河是工业区的港口。博埃里工作室的总体规划计划在正对着水面的地方建设一座新公园，并沿着岸边建设一系列高楼。这些建筑虽然目前属于港区的管辖范围，但也会对公众开放。鉴于未来发展趋势的不确定性，设计师设计了一栋双面的住宅综合楼，在遵循现有城市结构的同时也时刻准备着对水岸开放，将来甚至可能转变成一条海滨长廊。

The different heights of the blocks are related to the long views toward the inner city and to the solar orientation of the complex. While the north elevations of the buildings are treated in a rather "volumetric" way, the south ones are marked by the long horizontal lines of the overhanging balconies.

The whole complex is conceived and realized following the latest criteria for "sustainable" and energy conscious buildings. The building masses are carefully studied in relationship with the sun orientation, with an in-depth study of the shadow pattern at all hours and seasons both on the building surfaces and on the open collective spaces. The higher building is located on the north side and the lower one on the south side. The rows of balconies on the south side of the buildings screen the living rooms from the summer sun rays, while admitting the lower winter ones, greatly contributing to the energy efficiency of the complex. The north façades are marked by smaller openings contributing to low thermal transmittance. A significant part of the energy required by the building is provided by solar panels placed on the rooftop terraces of the two buildings.

不同高度的体量能欣赏到市区的远景，并考虑了太阳方位的影响。北立面被处理成整体结构，而南立面则分布了水平凸出的长阳台。

整栋建筑构想实现了下列的"可持续"和节能的最新标准。设计师认真研究了建筑体量与太阳方向，以及立面图案在全天候和各个季节时对建筑表面和开放公共空间的影响。较高的体量位于北侧，较矮的则位于南侧。位于建筑南立面的长阳台为客厅遮挡夏日的阳光辐射，到了冬天又能让阳光照入室内，极大地降低了公寓的能源消耗。北立面上的小窗户具有低导热率。另外屋顶上的太阳能电池板也对建筑的节能起着重要作用。

Location Plan
区位示意图

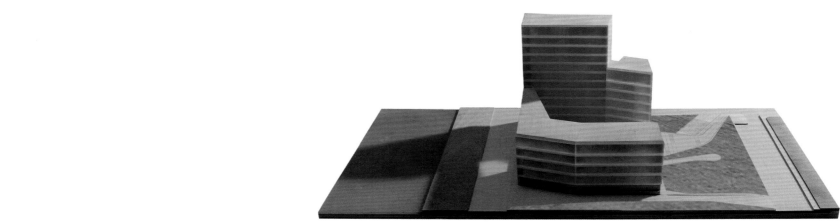

Model 1
模型图 1

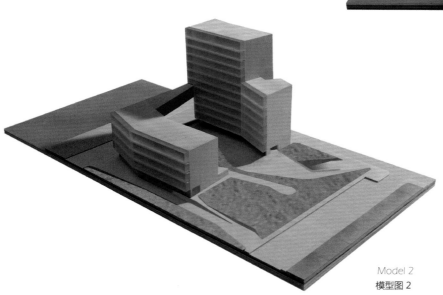

Model 2
模型图 2

Sketch1
草图1

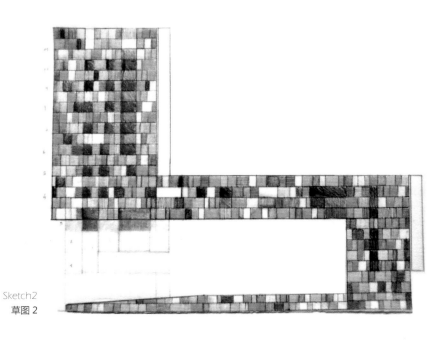

Sketch2
草图2

On the city side, a green rampart hosting the covered parking leads to a raised central court overlooking the water. A number of small shops and the atriums leading to the vertical distribution shafts open onto this semi-public "piazza", which will be connected to the water promenade by a ramp running parallel to the structure. The geometric inflections of the two building blocks and the lived-in "bridge" connecting them on the water side contribute to give a sense of spatial enclosure and intimacy to the central court.

建筑面对城市的一侧有绿色树屏穿过有顶棚的停车场，将人们引导至高起的中庭，俯瞰水上美景。许多小商店和中庭都连接到垂直分布轴，向着这个半公开的"广场"开放，并有一条和建筑结构平行的坡道连接到滨水长廊。两栋建筑的几何构造和连接它们的"桥梁"朝向水边，为中庭带来空间的围合感和亲密感。

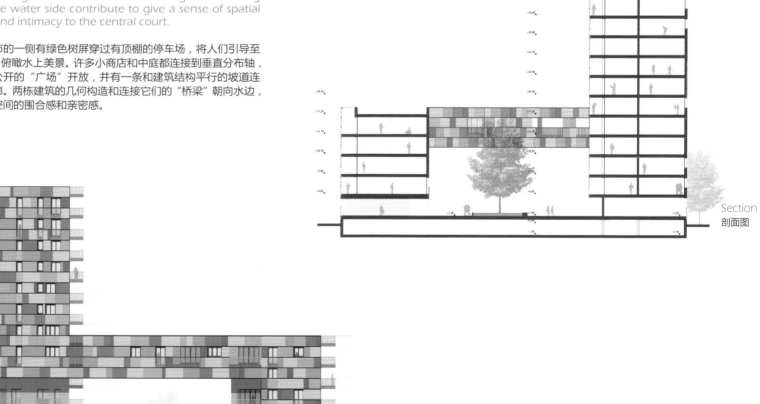

Section
剖面图

Elevation
立面图

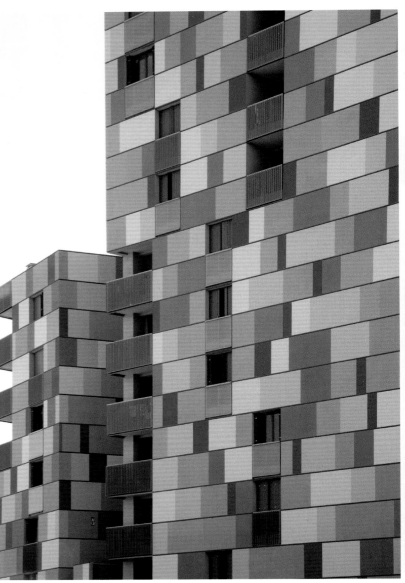

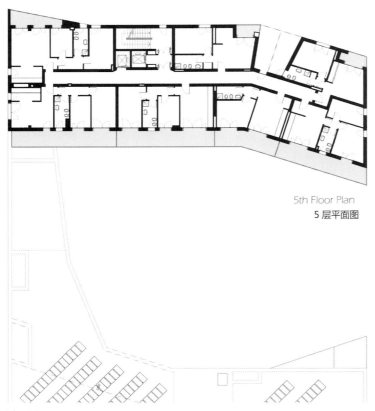

5th Floor Plan
5 层平面图

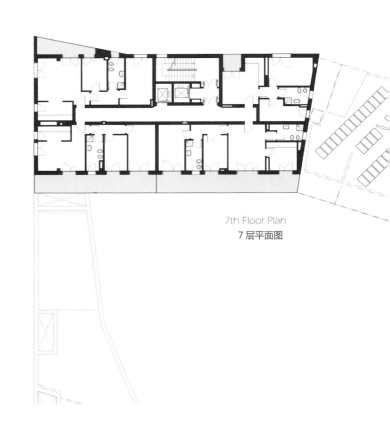

7th Floor Plan
7 层平面图

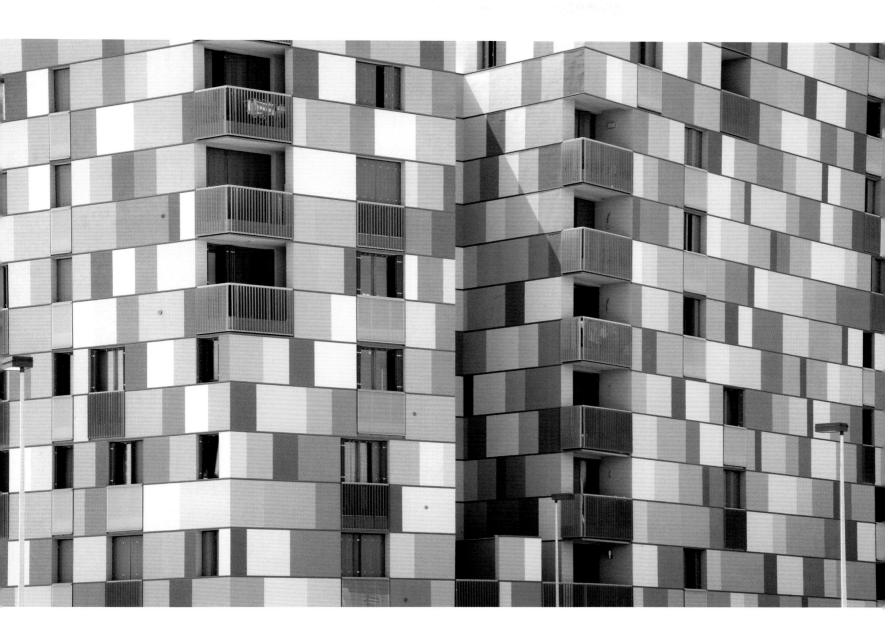

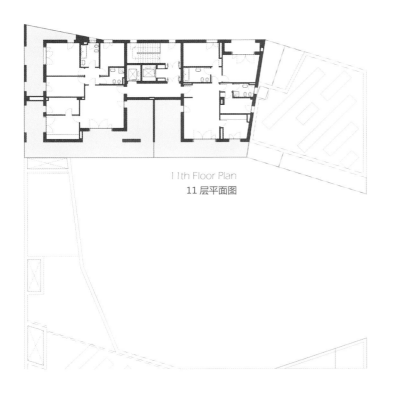

11th Floor Plan
11 层平面图

COLOR

The main façades of the building are marked by a number of terracotta horizontal "notched in" mouldings — two for every floor height — framing a plaster rendering of different shades of warm, clay-coloured shades in different hues and a cobalt blue one, creating a "mosaic" pattern — somehow inspired by Ravenna's famous Byzantine art — which generates a sort of scalar distortion in the perception of the building. This effect of momentary "camouflage" of the dimension of the building helps connecting its "domestic" dimension to its perception from the waterside and the city, where it stands alone as a temporary "landmark" waiting for the development to transform the landscape or this part of the city.

颜色

建筑的主立面镶嵌了许多水平的陶瓦方块——每两片组合在一起就有一层楼高。不同色调的暖色、土黄色和钴蓝色等，造就了一片"马赛克"图案的粉刷墙面。这种设计部分受到拉文纳著名的拜占庭艺术的启发，试图为建筑带来柔和的整体效果。这种视觉效果只是一种"障眼法"，将建筑的内部和对着水边和城市开放的外部联系起来。建筑将作为临时的"地标"，等待着这里的景观改造或是城市发展。

BALCONY 阳台
COLOR 色彩
LAYOUT 布局
MATERIAL 材料

De Entrée, Alkmaar, the Netherlands
荷兰阿尔克马尔 De Entrée 大厦

Architect: Arons en Gelauff Architecten
Location: Alkmaar, the Netherlands
Site Area: 13,000 m²
Floors: 10
Photography: Luuk Kramer, Arons en Gelauff Architecten

设计公司：Arons en Gelauff Architecten
地点：荷兰阿尔克马尔市
占地面积：13 000 平方米
层数：10
摄影：Luuk Kramer, Arons en Gelauff Architecten

The Overdie District in the renowned "cheese city "of Alkmaar is being restructured. This project is to renew the plaza, the heart of the neighborhood. The urban plan of Kraaijvanger-Urbis defines the new square with some grand gestures. The building on the west side is to be maintained. To the east and to the north, existing buildings, including a church, are demolished. Two major projects with some height accents come in its place. Phase one is the building "De Entrée", on the north side of the new square.

Overdie 区位于荷兰有着 "奶酪之城" 之称的阿尔克马尔市，而坐落在此的本项目旨在让作为社区中心的广场重新焕发活力。Kraaijvanger-Urbis 城市规划为新广场设定了几座建筑的位置：西侧的建筑保留，东侧和北侧原有的建筑，包括教堂在内都被拆除，取而代之是两栋具有一定高度的大型项目。一期工程就是本案的 De Entrée 大厦，位于新广场的南侧。

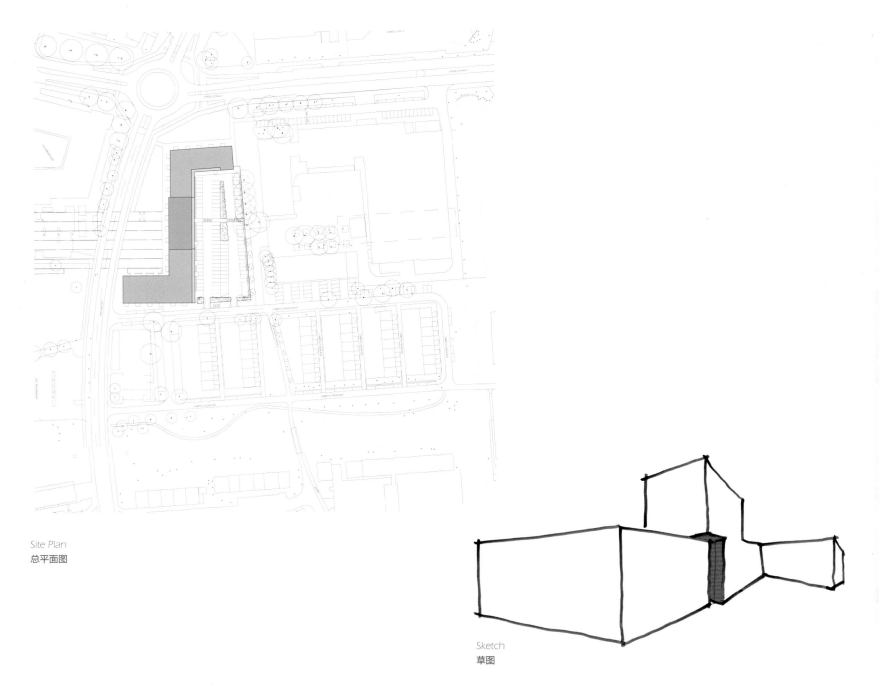

Site Plan
总平面图

Sketch
草图

The program for this building is a mix of affordable market apartments and social housing, including assisted living facilities. The ground floor houses commercial spaces and in the eastern flank on the lower two layers is a health center. The architecture strives for an optimistic approach that fits the postwar buildings. Parking is at ground level, behind the new building.

The situation is clear: a parking lot on the north side, the lively Geert Grote Square on the south of the building. Combined with the program for accessible housing, this determines the choice for the typology: a central core in the high part of the building with galleries that are included within the building volume.

In relation to the great height of the building, the given size of the high building part is not sufficient to really be able to make a tower. We make an asymmetric cut below the upper building element. It is this gesture that gives the building a unique form and offers a good marking of the entrance. The façades of each part of the building are designed with little differences in rhythm and detail, giving each side of the building its own character. Special attention is paid to the ambivalent character of the façade on the side of the main square. This formal side of the square is at the same time the sunny side for the apartments. The spacious balconies with colored glass soften the transition between the public and private space. These colors return in a modest way on the gallery side of the project. The car park will be planted.

大楼里主要有经济适用房和保障性住宅，及配套的援助性设施。大楼首层是商用店铺，东面的一、二层是保健中心。设计师希望项目以积极乐观的姿态作为战后建筑的象征。建筑后方的地下空间是停车场。

建筑的布局清晰：停车场位于北部，热闹的海尔特格罗特广场位于南部。住房出入便利的问题决定了建筑的造型：建筑高层的中央核心部分分布着突出的阳台，形成了建筑体量的主要特征。

考虑到建筑的高度，高出的部分有所限制，不能真正作为独立的塔楼。所以设计师在高出部分的下方设计了一个对称的切割口，如此一来赋予了建筑独特的造型和鲜艳的入口处。大楼外表造型统一，只在细节上稍有差异，使不同区域得以区分。面对主广场的立面特征鲜明，特别引人注目。这面既是广场的正面，也是住宅接受阳光的一面。大楼里的每间公寓都带有宽敞的阳台，窗户上安装了彩色玻璃，柔化了公共和私人空间的过渡。停车场也将种满植被。

Model 1
模型图 1

Model 2
模型图 2

FEATURE 特点分析

COLOR

The protruding red balconies are icons of the building. With red plates enclosed by dark red glass rails, the balconies become small paradises in plays of light and dark and richer level sense for residents enjoying sunshine and privacy. In addition, the bright red screens set on one side of the windows complement the red balconies nicely.

色彩

每层楼上凸出的亮红色阳台是建筑的标志物，鲜红色的地板由暗红色的玻璃栏杆围合起来，营造出一个个明暗、层次分明的小型空间，既能享受阳光，又能保护住户的隐私。此外，每户窗口的侧边还设置了鲜红色的屏风，和红色的阳台相得益彰。

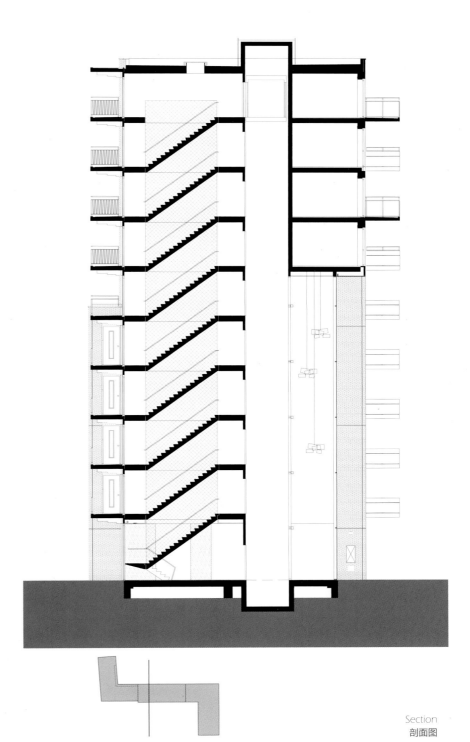

Section
剖面图

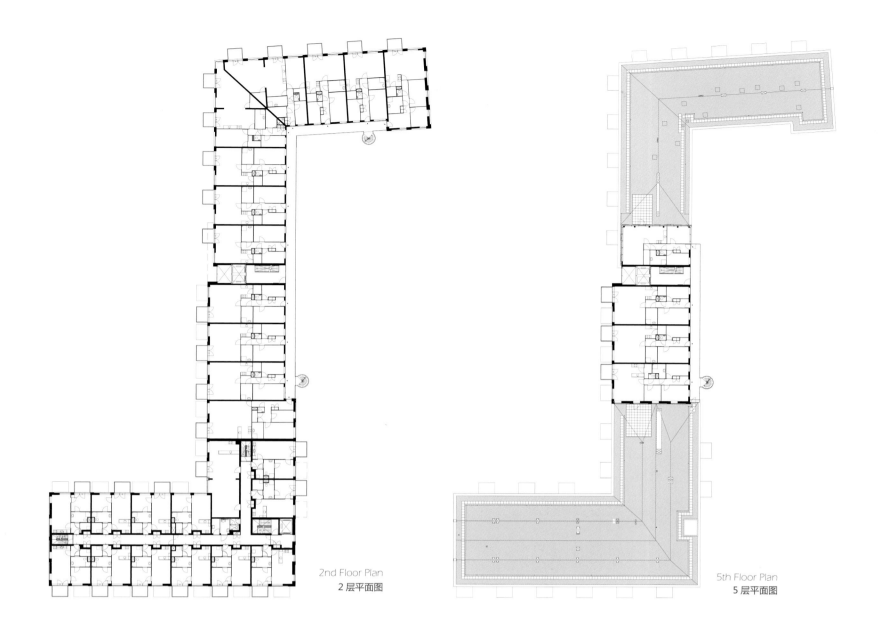

2nd Floor Plan
2 层平面图

5th Floor Plan
5 层平面图

STRUCTURE 结构

FAÇADE 立面

MATERIAL 材料

VIEW 视野

Miracle Residence, Istanbul, Turkey

土耳其伊斯坦布尔奇迹住宅

Architect: BFTA Mimarlık Ltd.
Client: Mön İnşaat Ltd.
Location: Istanbul, Turkey
Site Area: 17,000 m²
Gross Floor Area: 30,000 m²
Floors: 9
Height: 28.8 m
Photography: Gürkan Akay

设计公司：BFTA Mimarlık Ltd.
客户：Mön İnşaat Ltd.
地点：土耳其伊斯坦布尔市
占地面积：17 000 平方米
建筑面积：30 000 平方米
层数：9
高度：28.8 米
摄影：Gürkan Akay

The project is located at the Istanbul's Asian side near Sabiha Gokcen Int. Airport. Site is some 50 m.x300 m. in dimensions, along southeast-northwest direction and facing the highway at north.

本项目位于伊斯坦布尔临近亚洲一侧的地区，靠近 Sabiha Gokcen Int. 机场。场地大小约是 50 米 x 300 米，沿东南 - 西北朝向设计，北面是高速公路。

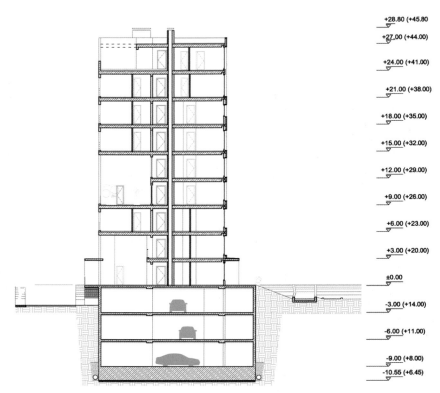

Cross Section 1
横截面图 1

The recently opened Sabiha Gokcen Int Airport is the only alternative in the city for the overcrowded European side Ataturk Int. Airport. Since the availability of the new airport, new demands for dwelling and retail promote denser land use and offer new opportunities for the Kurtkoy District.

Miracle Residence seems as a gate for this promising area. It consists two 100 m long blocks creating 250 m-long hotel and residential block with the social facilities between them. This 25 m x250 m surface is not designed as if its

another ordinary building in the area but, taken as a residential block design experiment.

Unique fibe-cement blocks, compact wooden lamine panels, standardized windows create its own harmony as design elements.

Since the block resides south edge of the site, there is so many potential uses on paysage, like waterscapes, sports grounds and pools.

原本靠近欧洲一侧的 Ataturk Int. 机场非常拥挤，最近新开 Sabiha Gokcen Int 机场为市民提供了另外一种选择。新机场投入使用后，附近对住宅和零售的需求让这里的土地使用量大幅提升，这也为 Kurtkoy 区创造了新的机遇。

本项目是这块新兴土地的大门，包括两栋长 100 米的建筑，中间加上酒店、住宅楼和其

他社区设施，总长 250 米。建筑占地 5 米 x 250 米，这座建筑不仅仅是一个普通的住宅项目，而且是一次建筑设计实验。

独特的体量，紧凑的木质板面和标准的窗户都是设计的元素。

因为建筑位于场地的南边，所以拥有大片的乡村风景，如水景、运动场和池塘等。

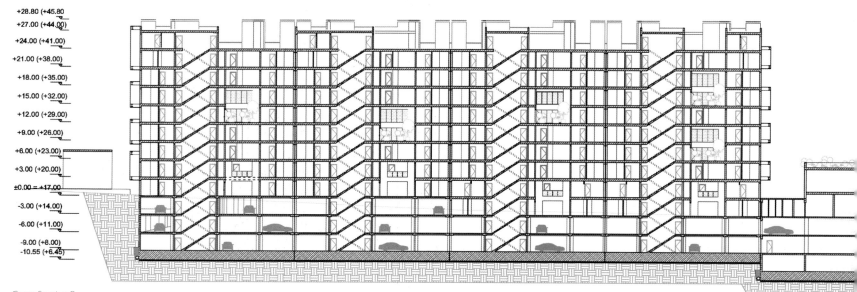

Cross Section 2
横截面图 2

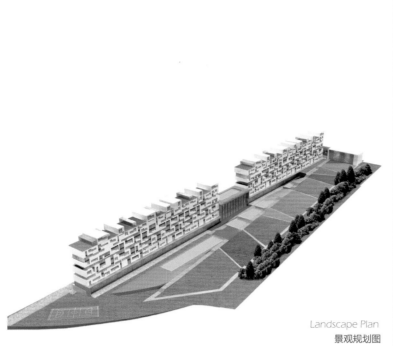

Landscape Plan
景观规划图

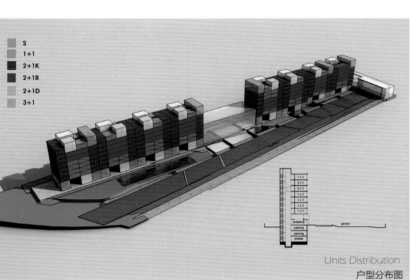

S
1+1
2+1K
2+1B
2+1D
3+1

1+1
2+1
2+1
1+1
1+1
1+1
1+1

entrance
parking
parking
shelter

parcan

Units Distribution
户型分布图

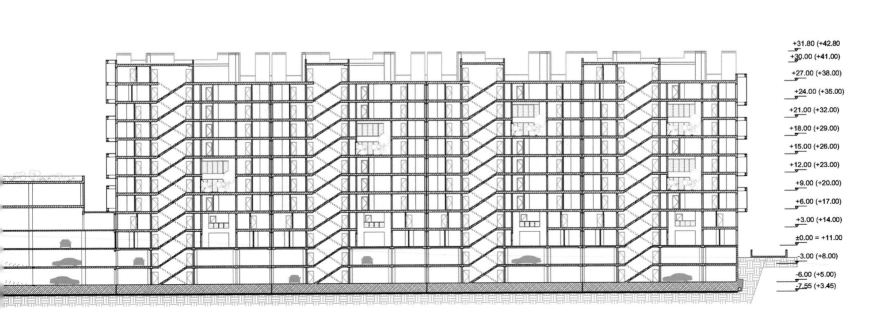

+31.80 (+42.80)
+30.00 (+41.00)
+27.00 (+38.00)
+24.00 (+35.00)
+21.00 (+32.00)
+18.00 (+29.00)
+15.00 (+26.00)
+12.00 (+23.00)
+9.00 (+20.00)
+6.00 (+17.00)
+3.00 (+14.00)
±0.00 = +11.00
-3.00 (+8.00)
-6.00 (+5.00)
-7.55 (+3.45)

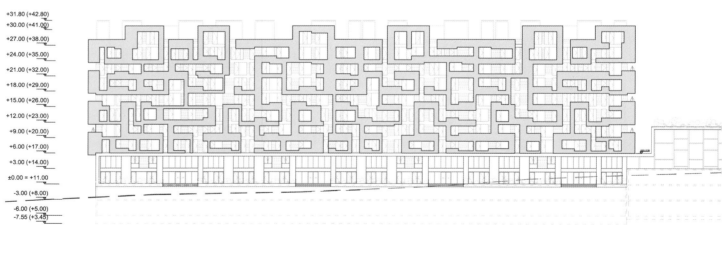

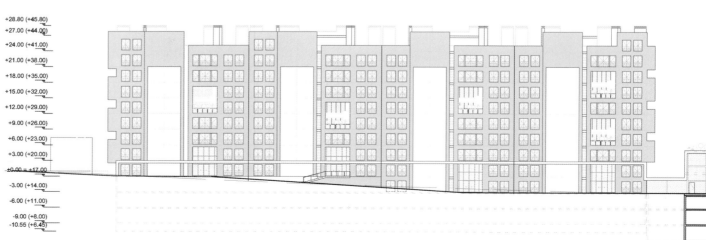

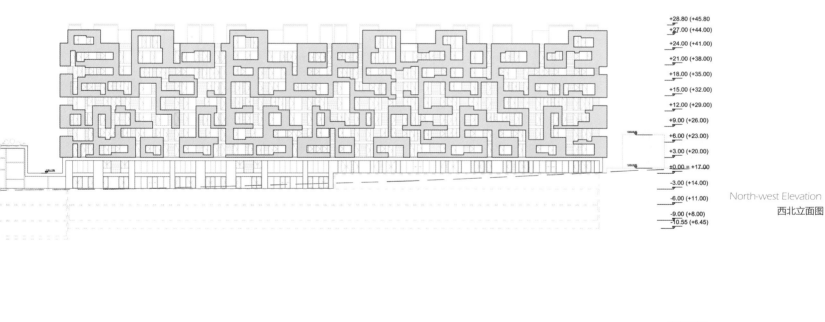

+28.80 (+45.80)
+27.00 (+44.00)
+24.00 (+41.00)
+21.00 (+38.00)
+18.00 (+35.00)
+15.00 (+32.00)
+12.00 (+29.00)
+9.00 (+26.00)
+6.00 (+23.00)
+3.00 (+20.00)
±0.00 = +17.00
-3.00 (+14.00)
-6.00 (+11.00)
-9.00 (+8.00)
-10.55 (+6.45)

North-west Elevation
西北立面图

+31.80 (+42.80)
+30.00 (+41.00)
+27.00 (+38.00)
+24.00 (+35.00)
+21.00 (+32.00)
+18.00 (+29.00)
+15.00 (+26.00)
+12.00 (+23.00)
+9.00 (+20.00)
+6.00 (+17.00)
+3.00 (+14.00)
±0.00 = +11.00
-3.00 (+8.00)
-6.00 (+5.00)
-7.55 (+3.45)

South-east Elevation
东南立面图

FAÇADE

As the gate of the promising area, the project featured in a simple shape and distinctive façade. Modular lines in white cover the entire façade as a maze interspersed with horizontal black lines and vertical brown lines. These lines bring rich level sense and linellae to the façade, showing infinite vitality with the emerging area.

立面

作为新兴地区的代表性建筑，本案以简洁的造型示人，而立面则展示出与众不同的一面。白色的模块化曲线如迷宫般布满整个建筑，间接穿插水平的黑色线条和垂直的棕色线条，让立面的层次感和线条更为丰富，与新地区一起展示出无限活力。

Tipical Floor Plan
标准楼层图

1+1 Apartment
1 + 1 公寓设计图

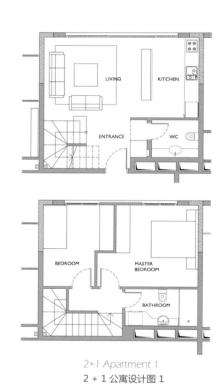

2+1 Apartment 1
2 + 1 公寓设计图 1

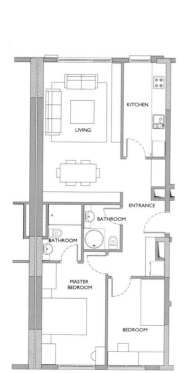

2+1 Apartment 2
2 + 1 公寓设计图 2

2+1 Apartment 3
2 + 1 公寓设计图 3

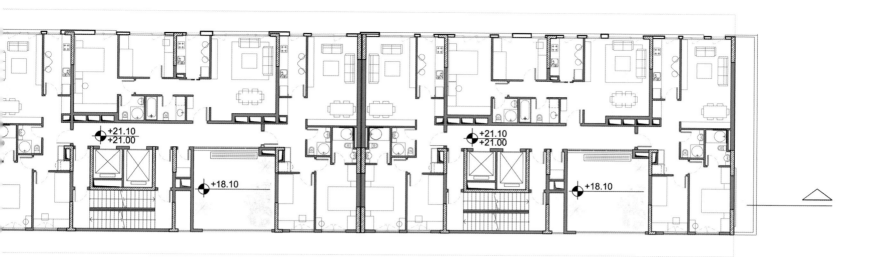

+21.10
+21.00

+18.10

+21.10
+21.00

+18.10

KITCHEN

DINING ROOM

LIVING

ENTRANCE

LAUNDRY

BATHROOM

BATHROOM

BEDROOM

BEDROOM

MASTER BEDROOM

3+1 Duplex Apartment
3 + 1 复式公寓设计图

BATHROOM

KITCHEN

BEDROOM

LIVING

Studio Plan
工作室平面图

MATERIAL 材料
FAÇADE 立面
SHAPE 造型
BALCONY 阳台

Milanofiori Residential Complex, Milano, Italy

意大利米兰米兰菲奥瑞综合住宅楼

Architect: Open Building Research S.r.l.
Client: Milanofiori 2000 S.r.l. (Brioschi Sviluppo Immobiliare Group)
Location: Milano, Italy
Gross Floor Area: 15,080 m²
Floors: 5
Photography: Apollonio Milanofiori, Nastasi Milanofiori

设计公司：开放建筑研究事务所
客户：米兰菲奥瑞 2000 S.r.l. (Brioschi 房地产开发集团)
地点：意大利米兰市
建筑面积：15 080 平方米
层数：5
摄影：Apollonio Milanofiori, Nastasi Milanofiori

The masterplan of Milanofiori is characterized by a series of functions (offices, hotels, restaurants, cinemas, leisure spaces, residences) that define together a cluster whose elements appear to follow the characteristics of the surrounding landscape, creating a public park as the extension of the existing forest. The design of the residential complex seeks the symbiosis between architecture and landscape, so that the synthesis of artificial and natural elements could define the quality of living and the sense of belonging by the inhabitants.

米兰菲奥瑞综合住宅楼总体规划的特色是功能丰富（写字楼、酒店、餐馆、影院、休闲空间、住宅），在由此形成的建筑组团中，每个元素似乎都遵循着周围的环境特点，从而打造了一座公园，成为现有森林的延伸。设计寻求建筑与景观的共生，综合人工与自然元素，从而能够界定生活品质和居民的归属感。

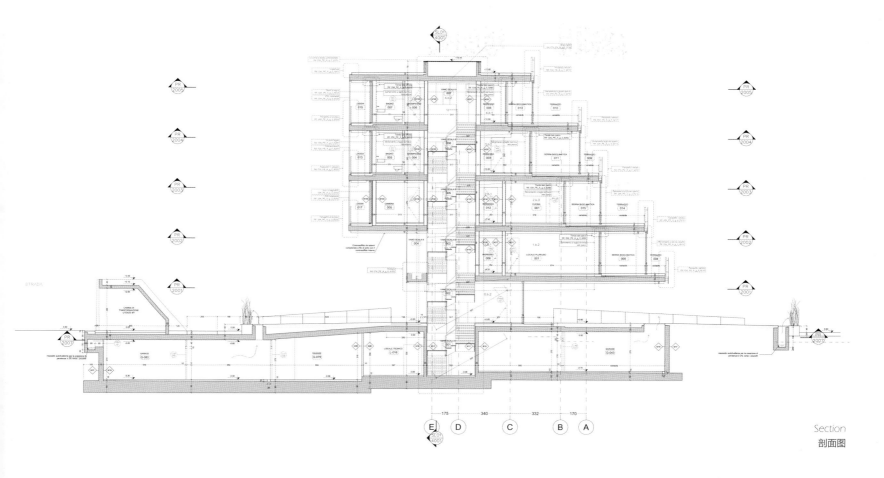

Section
剖面图

inside

inside - outside

outside

The interface between the building and the garden becomes the field where interaction between man and environment takes place. This interface is defined by the "C" form of the complex which encompasses the public park, and by the porosity from interior to exterior that characterizes all 107 apartments.

建筑与花园之间的界面成为人与环境互动的场所。这个界面被界定为 C 形的综合住宅群，它将公园紧密围绕。住宅楼共有 107 间公寓，从内到外都保持了通透的设计风格。

Inside - Outside Transition
室内外过渡分析图

FAÇADE

The two façades are designed differently: the one facing the street outside is more urban, and the one towards the inner park is more organic. The design of the urban façade stimulates a sense of belonging thanks to the composition of white frames which identify separately the units. These frames include vertical wooden panels of different widths which can slide across the frames and control the inner light as necessary.

立面

两个立面截然不同，一个面向街道，更为城市化；而另一个则面对着内部公园，更加生机盎然。城市化立面的设计由白色框架构成，营造出一种归属感，而且还能划分出各个住宅单元。框架由不同宽度的木制面板组成，这些面板不仅可沿着框架滑动，还可以在必要时控制内部的光线。

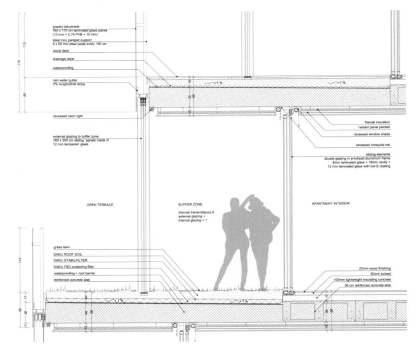

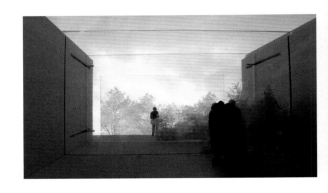

Detail 1
节点图 1

The geometry of the building is shaped by translation of the upper levels in line with positions of optimum solar exposure and by tapering of the external terraces in order to increase introspection among residents. The winter garden has a double value: an environmental value in providing a buffer zone which allows thermal regulation, and an architectural value in allowing extension of the interior living space towards the exterior landscape (and vice versa) permitting different uses from summer to winter.

Through the overlap of different natural layers (the public park, the open terraces and the winter gardens) the project seeks a kind of a holism of nature, where various personal interactions of these natural layers create an intensive landscape that is directly and personally customized by each resident.

In line with ever changing developments in contemporary living, the porosity of the architecture makes Milanofiori residential complex an evolving organism, in perpetual change, preferring the dynamic exchange between architecture and nature, and stimulating the interaction between man and environment.

建筑的几何形状是这样形成的：首先，上部楼层都是以接收到最佳太阳辐射的角度而设计的；其次，为了增加居民之间的交流机会，外部露台从上至下成阶梯状设计。冬季花园具有双重价值：从环境上看，这里可调节热能，成为一个缓冲区；从建筑学上看，它向外部景观延伸内部的起居空间（反之亦然）。无论在夏季还是在冬季，都有不同的用途。

通过不同自然层次（公园、开放露台和冬季花园）的重叠，该项目试图实现一种整体化的大自然层次，在这些自然层次中，各种人际互动关系创造了一种密集的景观，直接面向每一位居民，并为他们量身定制。

随着当代生活的不断变化与发展，通透的建筑设计风格使米兰菲奥瑞综合住宅楼成为了一个进化有机体。生命不息，变化不止，它更加注重建筑与自然之间的动态交换，同时激发人与环境之间的互动作用。

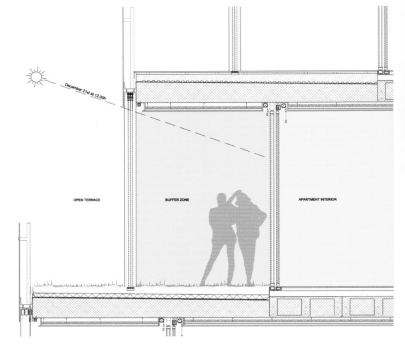

Detail 2
节点图 2

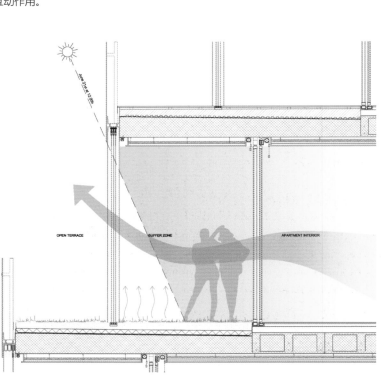

Detail 3
节点图 3

The organic façade overlooking the garden features double glazed bioclimatic greenhouses. The co-planarity between the glass of the greenhouse and the glass guardrail covering the string-course creates an effect where the shape of the construction and the background merge and reverse their roles constantly, producing kaleidoscopic effects overlapping the reflection of the public garden outside with the transparency of the private garden inside.

俯视着花园的生机盎然的立面采用双层玻璃窗，形成了符合生物气候学原理的温室。温室玻璃与玻璃护栏之间的共同平面在建筑外形与建筑相互融合的地方产生了一种效果，并不断地转换这两者的角色，将室外公共花园的映像与室内通透的私人花园重叠，给人带来千变万化的感受。

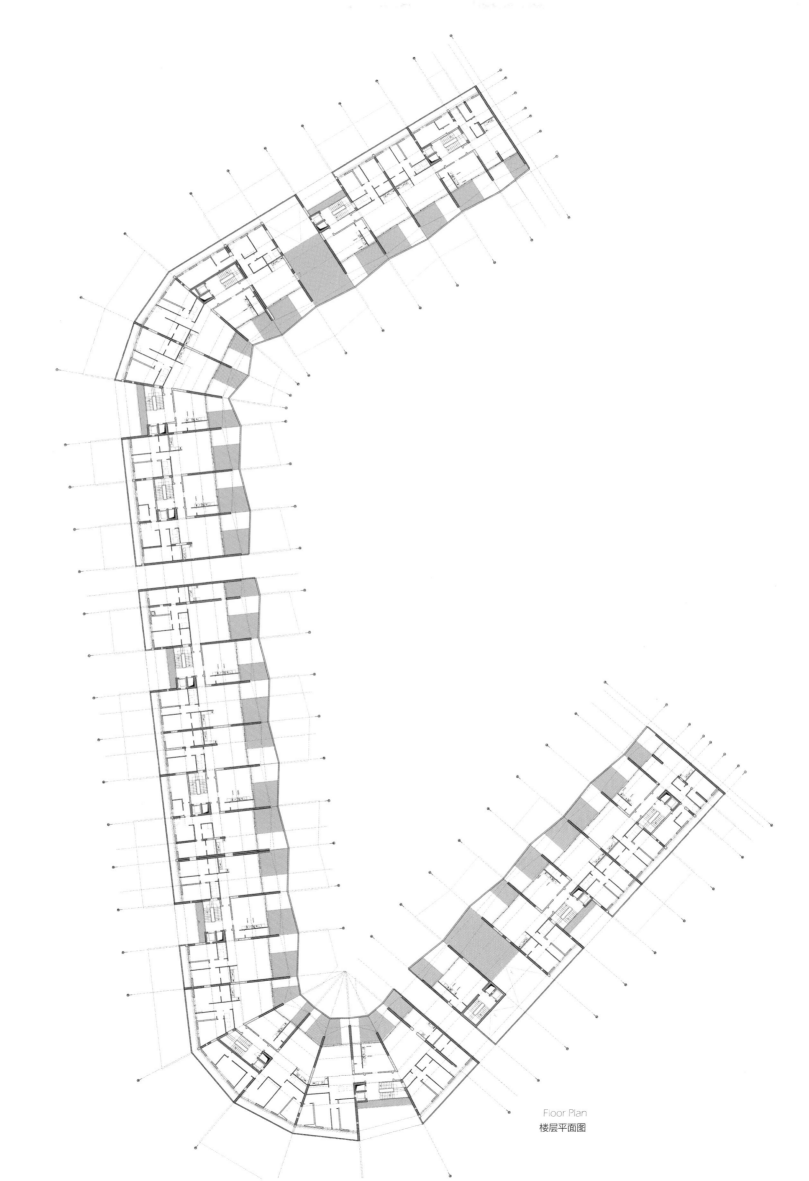

Floor Plan
楼层平面图

ENERGY-SAVING 节能

FAÇADE 立面

MATERIAL 材料

PATTERN 图案

Fremicourt, Paris, France

法国巴黎弗雷米古大厦

Architect: PERIPHERIQUES Architectes
Location: Paris, France
Site Area: 3,900 m²

设计公司：PERIPHERIQUES 建筑事务所
地点：法国巴黎市
占地面积：3 900 平方米

The plot allotted to the project is situated between Fremicourt Street and Boulevard de Grenelle. It is exceptional by its orientation and its center which are in continuity with the neighboring gardens. In order to achieve a low energy consumption building in Paris, it is fundamental to rely on important technical plans of action.

分配给该项目的建设地址坐落于弗雷米古街和德格勒内大道之间。它的朝向和中心都和邻近的花园连成一片。因此地理位置极为优越。同时，为了在巴黎实现低耗能建筑，实施关键的技术规划至关重要。

A PARIS

UN CHAUFFAGE PERFORMANT
+
UNE FACADE SUPER ISOLEE
+
AU SUD, UNE FACADE PHOTOVOLTAIQUE POUR PRODUIRE DE L'ENERGIE
+
UNE VENTILATION DOUBLE FLUX EFFICACE

▼

FACTURE ENERGETIQUE = 0

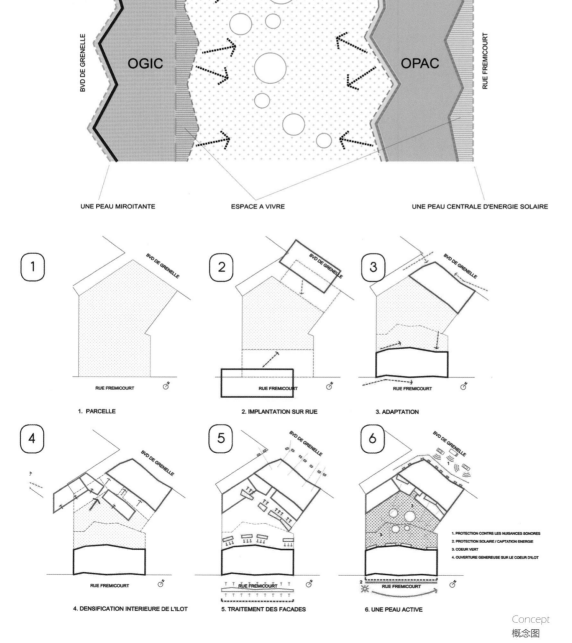

IMMEUBLE OGIC   COEUR D'ILOT VERT   IMMEUBLE OPAC

BVD DE GRENELLE

OGIC   OPAC

RUE FREMICOURT

UNE PEAU MIROITANTE   ESPACE A VIVRE   UNE PEAU CENTRALE D'ENERGIE SOLAIRE

FACADE EPAISSE IMMEUBLE OPAC
FACADE IMMEUBLE OGIG
FACADE VITREE
LAMES DE VERRE A CLAIRE VOIE
CLINS A CLAIRE VOIE

Concept
概念图

1. PARCELLE

2. IMPLANTATION SUR RUE

3. ADAPTATION

4. DENSIFICATION INTERIEURE DE L'ILOT

5. TRAITEMENT DES FACADES

6. UNE PEAU ACTIVE

1. PROTECTION CONTRE LES NUISANCES SONORES
2. PROTECTION SOLAIRE / CAPTATION ENERGIE
3. COEUR VERT
4. OUVERTURE GENEREUSE SUR LE COEUR D'ILOT

Concept
概念图

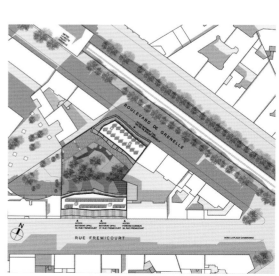

Site Plan
总平面图

The position of the building's body on the boulevard (nine levels high) and on the street (ten levels high) is planned in a way to allow housing units with a double exposition. Their south end is extended by loggia spaces with pleasant views. Beyond the general implantation question, the proposed working drawing of the building is adapted to the context's constrains.

建筑的一侧是林荫大道（9层高），而另一侧是宽广的街道（10层高），这就使得建筑能获得双重的采光。南面末端通过凉廊空间延伸，给人一种舒适的视觉感受。对于普遍的环境融入问题，建筑的设计理念就是适应周围环境的限制。

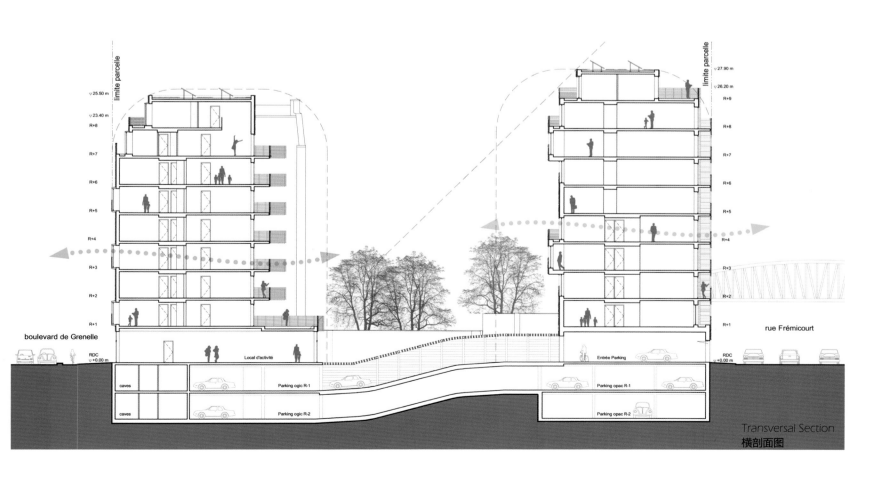

Transversal Section
横剖面图

South Elevation
南立面图

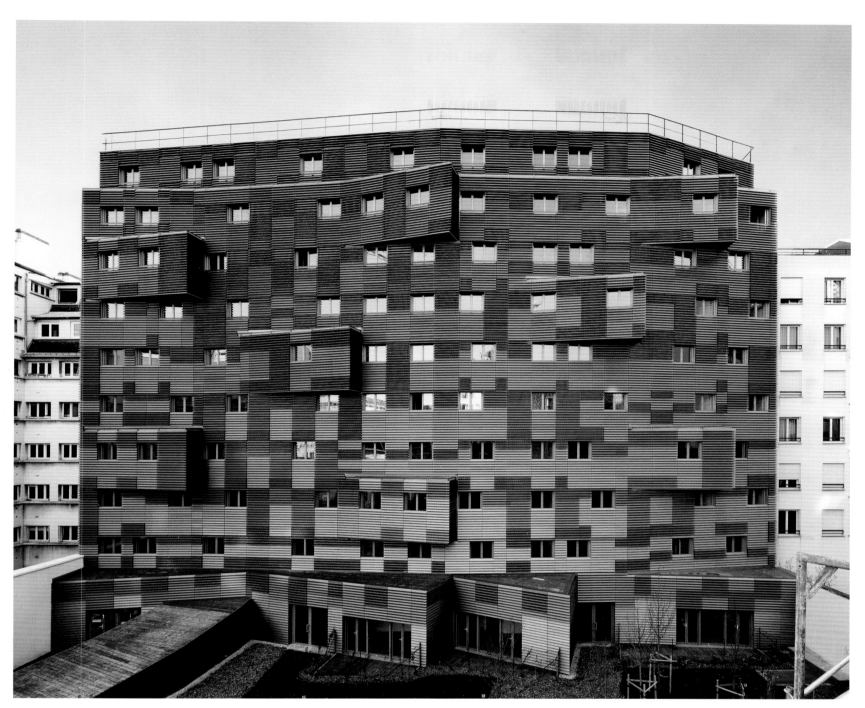

North Elevation
北立面图

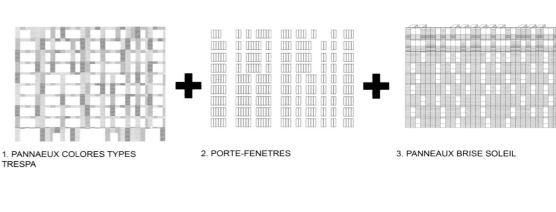

1. PANNAEUX COLORES TYPES TRESPA

2. PORTE-FENETRES

3. PANNEAUX BRISE SOLEIL

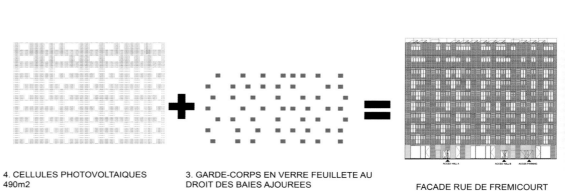

4. CELLULES PHOTOVOLTAIQUES 490m2

3. GARDE-CORPS EN VERRE FEUILLETE AU DROIT DES BAIES AJOUREES

FACADE RUE DE FREMICOURT

Façade Concept 1
立面概念图 1

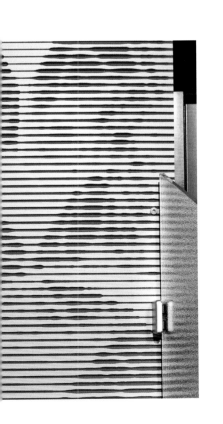

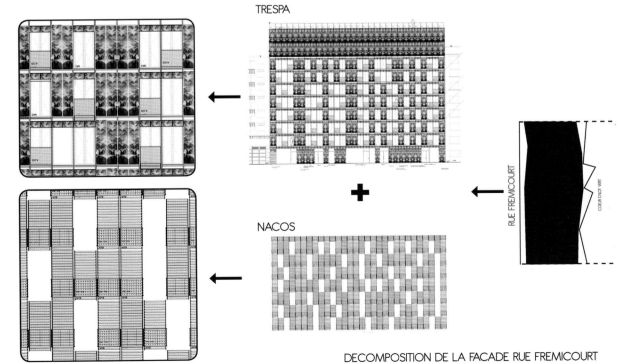

TRESPA

NACOS

DECOMPOSITION DE LA FACADE RUE FREMICOURT

Façade Concept 2
立面概念图 2

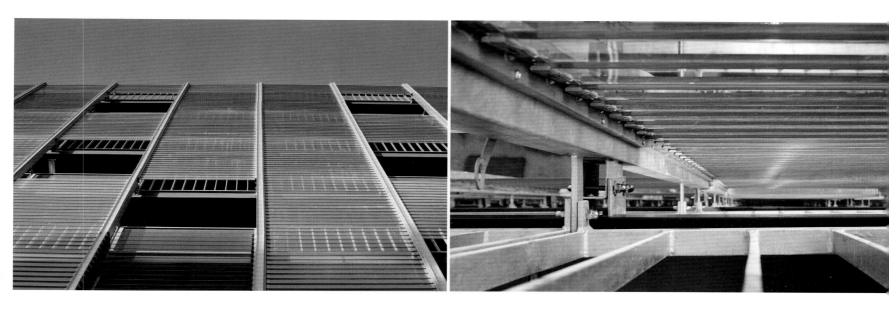

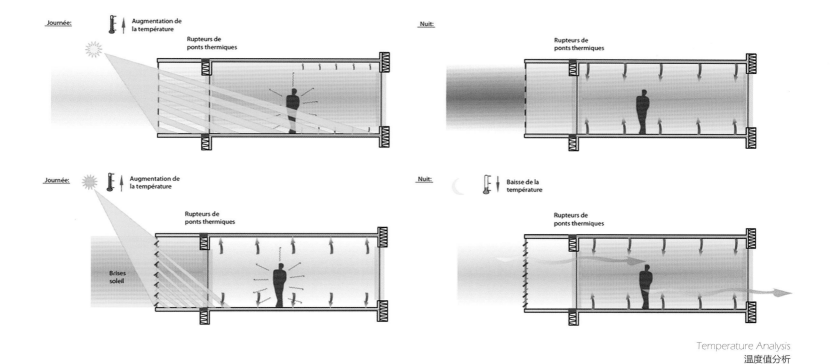

Journée: Augmentation de la température

Rupteurs de ponts thermiques

Nuit: Rupteurs de ponts thermiques

Journée: Augmentation de la température

Rupteurs de ponts thermiques

Brises soleil

Nuit: Baisse de la température

Rupteurs de ponts thermiques

Temperature Analysis
温度值分析

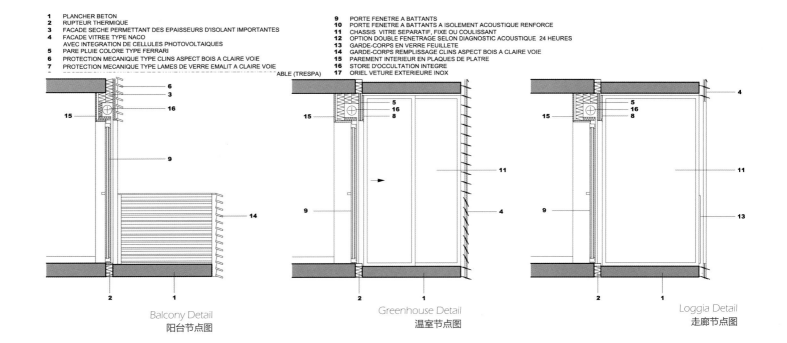

1 PLANCHER BETON
2 RUPTEUR THERMIQUE
3 FACADE SECHE PERMETTANT DES EPAISSEURS D'ISOLANT IMPORTANTES
4 FACADE VITREE TYPE NACO
   AVEC INTEGRATION DE CELLULES PHOTOVOLTAIQUES
5 PARE PLUIE COLORE TYPE FERRARI
6 PROTECTION MECANIQUE TYPE CLINS ASPECT BOIS A CLAIRE VOIE
7 PROTECTION MECANIQUE TYPE LAMES DE VERRE EMALIT A CLAIRE VOIE
   ABLE (TRESPA)

9 PORTE FENETRE A BATTANTS
10 PORTE FENETRE A BATTANTS A ISOLEMENT ACOUSTIQUE RENFORCE
11 CHASSIS VITRE SEPARATIF, FIXE OU COULISSANT
12 OPTION DOUBLE FENETRAGE SELON DIAGNOSTIC ACOUSTIQUE 24 HEURES
13 GARDE-CORPS EN VERRE FEUILLETE
14 GARDE-CORPS REMPLISSAGE CLINS ASPECT BOIS A CLAIRE VOIE
15 PAREMENT INTERIEUR EN PLAQUES DE PLATRE
16 STORE D'OCCULTATION INTEGRE
17 ORIEL VETURE EXTERIEURE INOX

Balcony Detail
阳台节点图

Greenhouse Detail
温室节点图

Loggia Detail
走廊节点图

The four façades, isolated on the interior, have been enveloped in an openwork horizontal sheathing elements — using glass for the Fremicourt side, anodized aluminium and wood for the garden side, and finally enamelled terracotta for the Boulevard de Grenelle side.

建筑四周的外立面独立于内部，被一层网状的水平外壳覆盖，面对着弗雷米古街一侧的墙面以玻璃为材质，而面对着花园一侧使用了混合阳极氧化铝和木质的材质，最后面对着德格勒内大道的墙面则是采用了赤土色的瓷漆。

FEATURE 特点分析

FAÇADE

The façades are creased in order to exploit at best its exposition and to mark the opening to the surrounding free spaces. The program's particularity is that the same operation unites social housing as well as private housing units. Thus, we have imagined to treat the façades in a common way but with some declension.

立面

建筑的立面设计成褶皱的形式，使采光更为充足，同时让空间显得更加通透。项目的独到之处在于将相同设置的房间分为统一社会住房和私人住宅。因此，设计师采用常见又独特的方式设计立面。

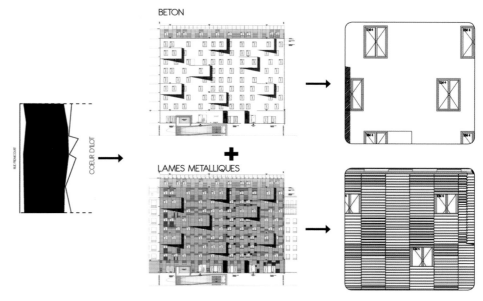

DECOMPOSITION DE LA FACADE COTE COEUR D'ILOT

Oriel Window Detail
飘窗节点

1	PLANCHER BETON	9	PORTE FENETRE A BATTANTS
2	RUPTEUR THERMIQUE	10	PORTE FENETRE A BATTANTS A ISOLEMENT ACOUSTIQUE RENFORCE
3	FACADE SECHE PERMETTANT DES EPAISSEURS D'ISOLANT IMPORTANTES	11	CHASSIS VITRE SEPARATIF, FIXE OU COULISSANT
4	FACADE VITREE TYPE NACO	12	OPTION DOUBLE FENETRAGE SELON DIAGNOSTIC ACOUSTIQUE 24 HEURES
	AVEC INTEGRATION DE CELLULES PHOTOVOLTAIQUES	13	GARDE-CORPS EN VERRE FEUILLETE
5	PARE PLUIE COLORE TYPE FERRARI	14	GARDE-CORPS REMPLISSAGE CLINS ASPECT BOIS A CLAIRE VOIE
6	PROTECTION MECANIQUE TYPE CLINS ASPECT BOIS A CLAIRE VOIE	15	PAREMENT INTERIEUR EN PLAQUES DE PLATRE
7	PROTECTION MECANIQUE TYPE LAMES DE VERRE EMALIT A CLAIRE VOIE	16	STORE D'OCCULTATION INTEGRE
8	PROTECTION MECANIQUE TYPE PANNEAUX DE RESINE THERMODURCISSABLE (TRESPA)	17	ORIEL VETURE EXTERIEURE INOX

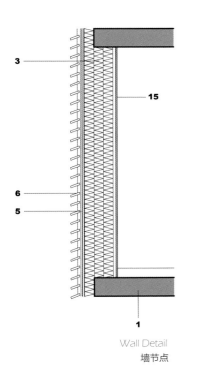

Wall Detail
墙节点

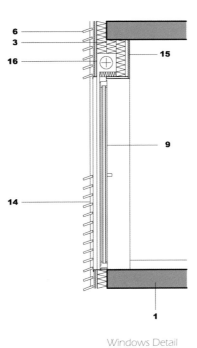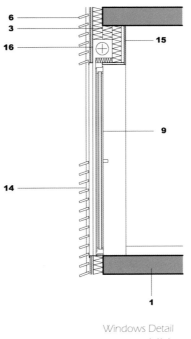

Windows Detail
窗节点

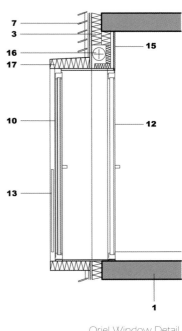

Oriel Window Detail
飘窗节点

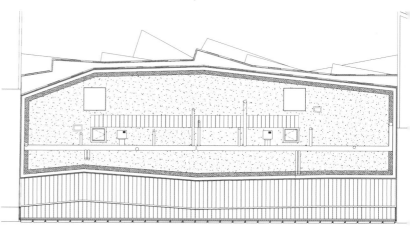

Roof Section
屋顶剖面图

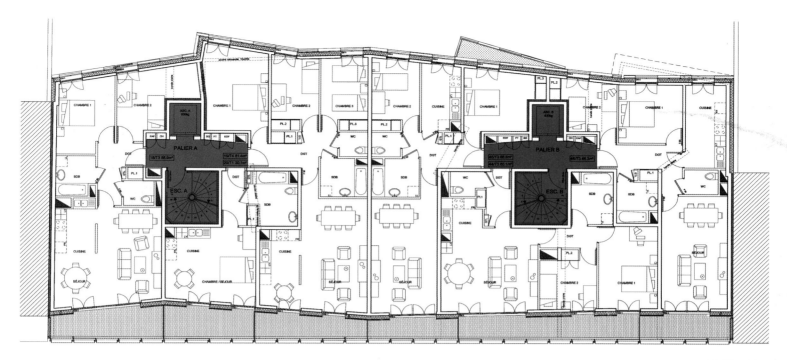

6th Floor Plan
6层平面图

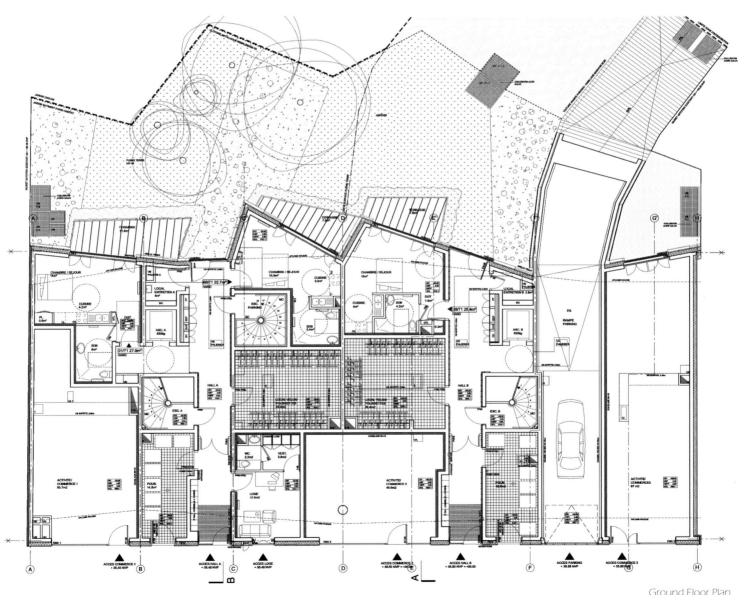

Ground Floor Plan
首层平面图

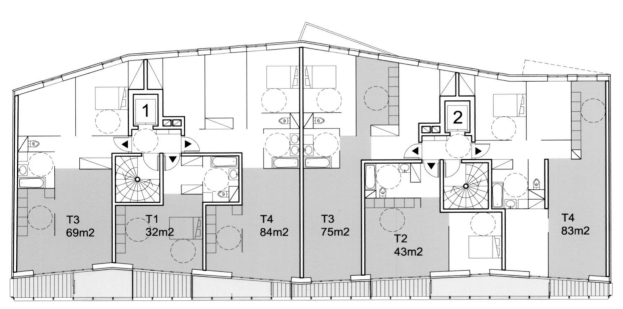

Floor Plan 1
楼层图 1

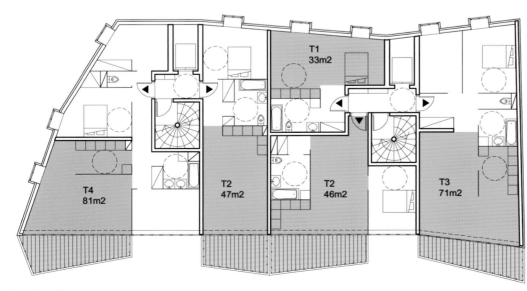

Floor Plan 2
楼层图 2

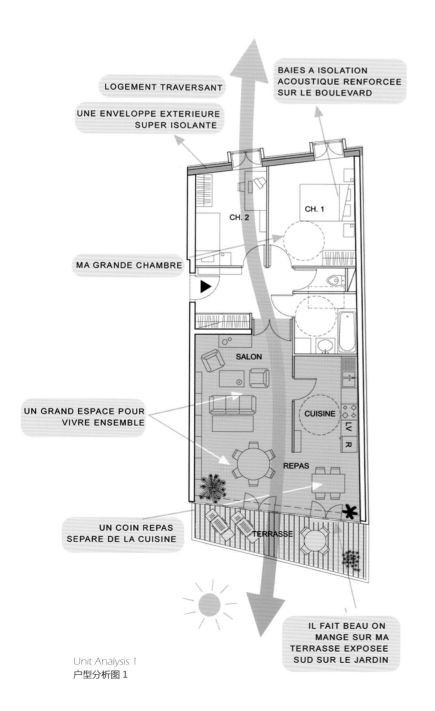

LOGEMENT TRAVERSANT

BAIES A ISOLATION
ACOUSTIQUE RENFORCEE
SUR LE BOULEVARD

UNE ENVELOPPE EXTERIEURE
SUPER ISOLANTE

MA GRANDE CHAMBRE

CH. 2

CH. 1

UN GRAND ESPACE POUR
VIVRE ENSEMBLE

SALON

CUISINE

LV

R

REPAS

UN COIN REPAS
SEPARE DE LA CUISINE

TERRASSE

IL FAIT BEAU ON
MANGE SUR MA
TERRASSE EXPOSEE
SUD SUR LE JARDIN

Unit Analysis 1
户型分析图 1

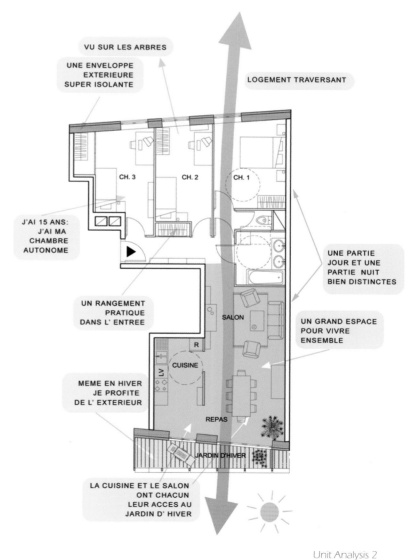

VU SUR LES ARBRES

UNE ENVELOPPE
EXTERIEURE
SUPER ISOLANTE

LOGEMENT TRAVERSANT

CH. 3

CH. 2

CH. 1

J'AI 15 ANS:
J'AI MA
CHAMBRE
AUTONOME

UNE PARTIE
JOUR ET UNE
PARTIE NUIT
BIEN DISTINCTES

UN RANGEMENT
PRATIQUE
DANS L' ENTREE

SALON

UN GRAND ESPACE
POUR VIVRE
ENSEMBLE

R

CUISINE

LV

MEME EN HIVER
JE PROFITE
DE L' EXTERIEUR

REPAS

JARDIN D'HIVER

LA CUISINE ET LE SALON
ONT CHACUN
LEUR ACCES AU
JARDIN D' HIVER

Unit Analysis 2
户型分析图 2

LAYOUT 布局
FAÇADE 立面
MATERIAL 材料
BALCONY 阳台

Grosvenor Waterside, London, UK
英国伦敦格罗夫纳河岸住宅

Architect: Make Architects, Sheppard Robson
Client: St James Group Ltd.
Location: London, UK
Site Area: 25,000 m²
Floors: 9
Photography: Cooper Rose, Z Olsen
Drawings: Make Architects

设计公司：Make Architects、Sheppard Robson
客户：圣雅各福群集团有限公司
地点：英国伦敦市
占地面积：25 000 平方米
层数：9
摄影：Cooper Rose、Z Olsen
图纸：Make Architects

Located at the heart of the restored Grosvenor Dock in west London, this residential project has transformed a derelict former industrial site into a new urban quarter.

这个去年开始施工的大型住宅工程位于新近重新开发的西伦敦格罗夫纳码头的中心，它已经把这个原来荒废多年的工厂区转变为活跃的新城区。

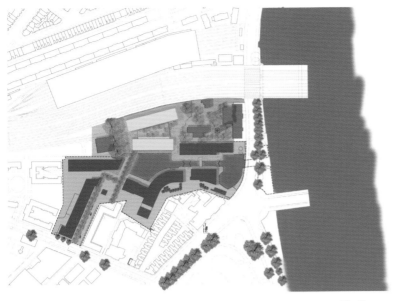

103 private and 196 affordable apartments are housed within two blocks of accommodation angled to maximize views out towards the River Thames. Each block is scaled to complement its surroundings: the larger of the two addresses the adjacent listed buildings of Churchill Gardens, while the smaller faces onto the dock and features a colonnade containing a restaurant, bar and entrance lobbies. The two blocks are linked at ground level by a single storey of commercial accommodation with a planted roof landscaped to provide an additional green terrace area for the residents.

两幢分别拥有 103 套私人公寓和 193 套经济适用房的建筑物，经过巧妙的朝向设计，使更多的住宅单元能够尽享泰晤士河的景观。建筑的体量设计，使其与周边建筑相互协调，两者中体量较大的与相邻的丘吉尔花园的文物保护建筑相呼应；较小且有着柱廊的建筑则面向码头，其清静的环境自然地成为餐厅、酒吧以及建筑入口大厅的所在。两幢建筑由地面一层的商业空间所联系，同时屋顶花园也为住户们提供了更多的绿化空间。

Site Plan
总平面图

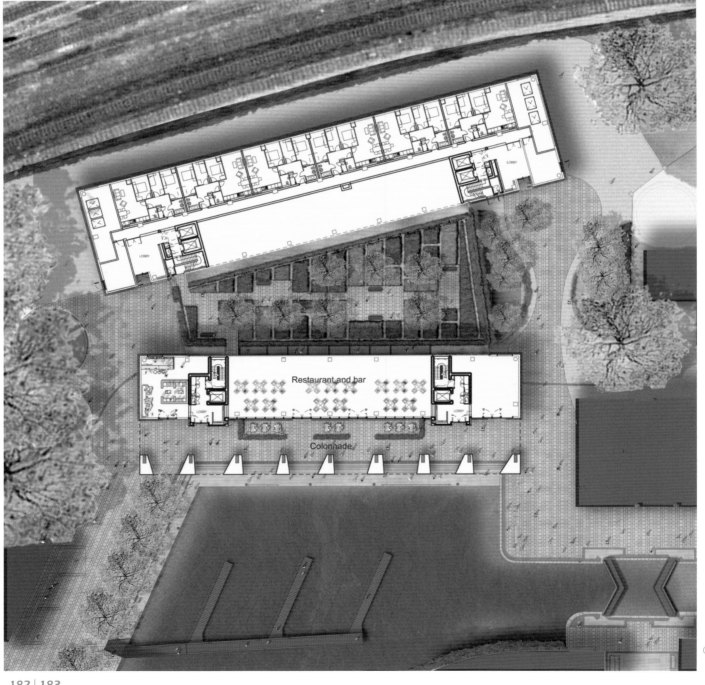

Ground Floor Plan
首层平面图

Elevation
立面图

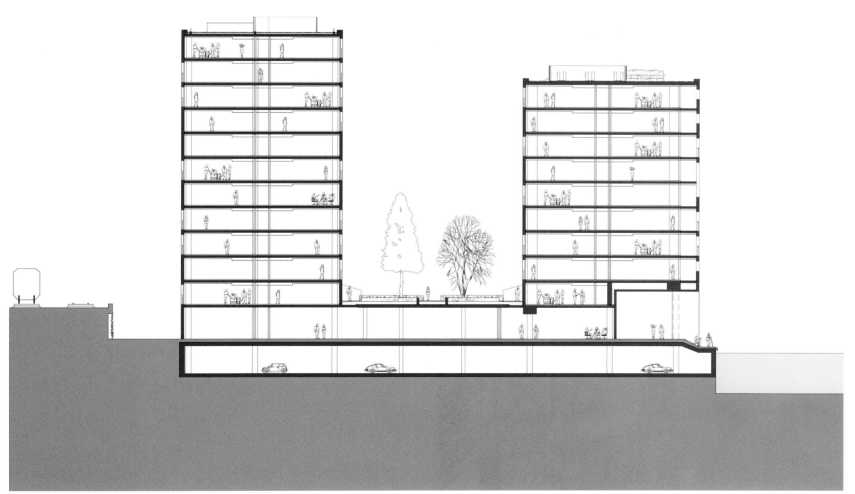

Section
剖面图

Sketch
草图

FEATURE 特点分析

FAÇADE

Generous enclosed balconies incise the façade with a rhythmic arrangement of voids that draw light and fresh air into living areas, while slender vertical and horizontal glazing slots create more intimate and subtle lighting for bedrooms and bathrooms.

The cladding system is etched in a design developed by artist Clare Woods, transforming the building into a canvas for a monumental public art work.

立面

半开放式的露台，将立面切割为有韵律的空间构成，并且使阳光和新鲜空气流泄至室内，细长的横向与竖向的玻璃窗为卧室和卫生间提供了柔和的光照。

此外，艺术家 Clare Woods 应邀设计的树木图案将由金属蚀刻技术浮刻至铝板幕墙上。

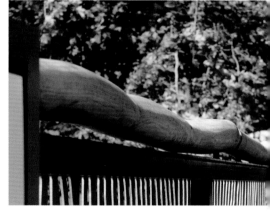

FAÇADE 立面

SHAPE 造型

OPEN SPACE 开放空间

LAYOUT 布局

Capella Apartments, Sydney, Australia

澳大利亚悉尼卡贝拉公寓

Architect: Francis -Jones Morehen Thorp
Location: Sydney NSW, Australia
Gross Floor Area: 18,750 m²
Floors: 5
Photography: John Gollings, Mark Donaldson

设计公司：Francis -Jones Morehen Thorp
地点：澳大利亚新南威尔士州悉尼市
建筑面积：18 750 平方米
层数：5
摄影：John Gollings, Mark Donaldson

This mixed-use project for 164 apartments has given us the opportunity to develop open space, urban design and architectural form ideas within an important and sensitive district of Sydney.

本项目具有多重功能，共有 164 套公寓。本案为设计师在悉尼的一个重要而敏感的地区大力发展开放空间、城市设计和建筑形态提供了机遇。

Sketch
草图

Site Plan
总平面图

SHAPE

Within the intersecting forms are a wide range of apartment types with varying section and plan forms. The street front form incorporates deeply recessed and protected terraces creating open space "rooms" while giving daylight to multiple interior spaces. The fully glazed suspended volumes include a continuous stepped section to create linear thermal chimneys within every room. The garden elevation of these glass forms incorporates a series of continuous vertical blades that are twisted to the north to gently adjust the orientation of the apartments towards the sun and view.

造型

公寓外部是交叉的形式，内部的公寓类型则多样化，面积和布局各不相同。临街面部分深深缩进，以保护大厅和创造开放的室内空间，保证室内各空间都光线充足。全玻璃立面上有一系列悬浮突出的部分，这是为每间公寓设置的直线形热烟囱。面向花园的立面上有一系列连续的垂直叶片元素朝向北方，用途是稍微调节建筑面对阳光和视野的朝向。

Formally the project is a sequence of intersecting layers and volumes: A landscaped podium forms a folding groundplane of open spaces and gardens while absorbing car park loading and back of house retail accommodation. "Resting" on this solid platform are two composed and articulated forms of tiled precast and frame that extend out to hold the line and scale of the street. Between podium and street form is the space of the retail and apartment foyers protected by a single awning plane that folds up to mirror the stepping in the platform and open this zone to the public street.

项目由一系列交叉的层面和体量组成：景观区域是一片折叠的开放空间，花园同时作为停车场，并为建筑背后的商店提供场所。两栋覆盖着框架结构预制砖瓦的建筑体量坐落在坚实的地基上，保持和街道边缘的方向一致。建筑和街道之间是商店区和公寓的大厅，大厅有遮阳篷，折叠起来后可以让大厅的楼梯光线更明亮，并让大厅空间与外部街道联系起来。

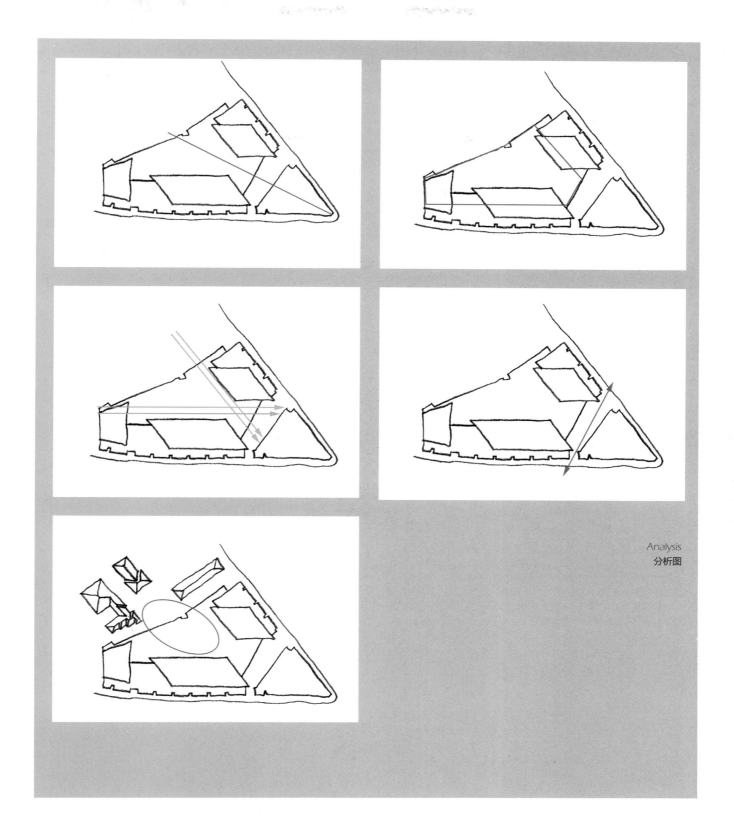

At the heart of this project is the creation of a new open space sequence that begins with the broadening of the footpaths through stepped terraces and setback shops and extends into a new landscaped public square that connects two existing streets. This new public square has been created on previously private land. It is orientated north to receive sunlight through the sharp angled forms of the new apartment volumes and is bordered by a stepped series of landscaped terraces that open into the shared communal gardens of the apartments. This communal landscaped open space of gently folding planes pool and trees follows a triangular geometry, is defined and embraced by the two wings of apartment volumes and opens wide to the northern sun.

This project has also given us the opportunity to continue to explore the elaboration of architectural form through the editing expression of construction techniques, systems and natural materials.

项目的中心是一片开阔空间，分布了宽阔的人行道，穿过台阶平台和后部的商店，来到新建的景观公共广场，最后连接到两条原有的街道。公共广场所在的地方原本是一片私人用地，朝向南方，阳光穿过公寓的锐角体量能照射到广场上。另有一系列台阶露台，在让广场显得更为宽阔的同时，也可作为公寓公共花园的一部分。这片公共花园里的水池和树木都呈折叠式分布，形成三角形布局；公寓建筑的两翼环绕、界定出花园的界限，让花园面向南方，接收更多阳光。

本项目也为设计师提供了通过对建筑技术、系统和天然材料的表达继续研究建筑形式的好机会。

Free from the scale and modulation of the street are two fully glazed distorted rectangular forms of precise geometry that "hover" and intersect the street forms. These suspended glass angled forms give the project a direction and energy towards the urban corner while creating a peaceful, simple and calm definition of the triangular communal gardens in contrast to the business of the street.

全玻璃立面的扭曲长方形体量并不会受到街道规模和管制的影响，其精妙的几何形态叠立在街道上，与其相交成为一体。看似悬浮的玻璃转角形式让建筑面对城市的一角更为直观、富有活力，并打造出一个平静简单的三角公共花园，与繁忙的街道形成对比。

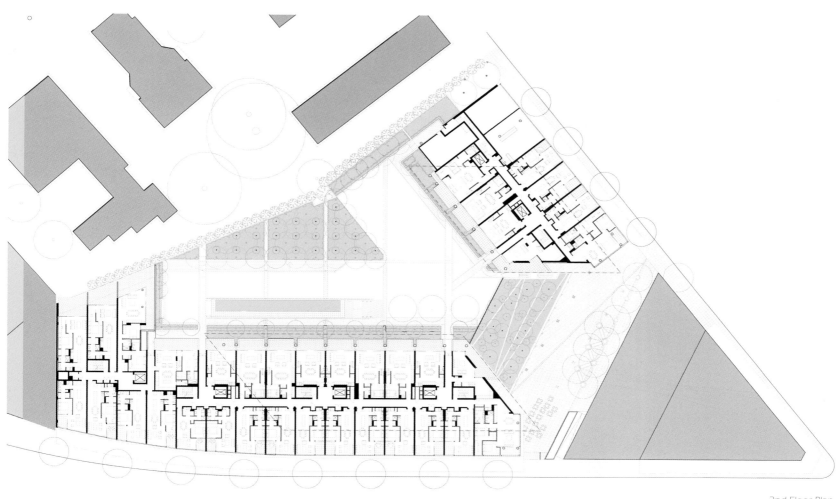

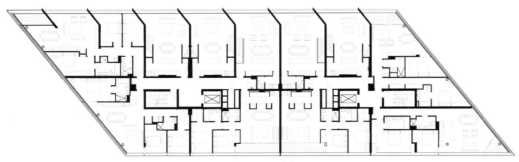

7th Floor Plan
7 层平面图

FAÇADE 立面
SUSTAINABILITY 可持续性
SHAPE 造型
COLOR 色彩

Little Bay, Australia
澳大利亚小湾公寓

Architect: Francis-Jones Morehen Thorp
Client: Stockland
Location: NSW Australia
Gross Floor Area: 5,835 m²
Floors: 5
Photography: John Gollings

设计公司：Francis-Jones Morehen Thorp
客户：斯多克兰房产公司
地点：澳大利亚新南威尔士州
建筑面积：5 835 平方米
层数：5
摄影：John Gollings

The design proposals for Manta, and its sister Alaris, have been drawn from an analysis and reading of the particular qualities and characteristics of the locality and context. Their sinuous forms compress and then open out in a strong entrance sequence to the Prince Henry site, reinforcing the corners at the intersection of Pine Avenue and Anzac Parade, then stepping down and peeling back to reveal a new public space and the elegant façades of the Flowers Wards, the heritage landscape and coastal vistas beyond. The new north facing open space and ground level retail accommodation create both a community and commercial heart for the Prince Henry Community.

设计师为 Manta 和 Alaris 而做出的设计方案是对当地环境中的特质和特点中进行解读和分析而得出的。建筑的造型蜿蜒紧凑，向着亨利王子社区打开宽敞的入口，在松木大道和安泽克大道的交叉点处转弯，最后下降并回转，围合出新的公共广场和花园的优雅立面，远处是文物景观和海景。新增设的北向开放空间和首层作为商店，成为了亨利王子社区的社交和商业中心。

Behind the screened façades, raised above street level, three to four levels of apartments are oriented toward the easterly views, the city and Botany Bay. A random pattern forms the primary street walls, while providing shade and privacy to generous private open space and interiors beyond. The orientation of the apartments and avoidance of overlooking are a direct result of building form and internal planning initiatives. In particular, the overlooking of primary living areas and external private open spaces, as well as the number of apartments facing each other across Pine Avenue, is minimized. External shading systems and staggered vertical blades further augment the performance of the buildings.

后面地基是高于街面 3 至 4 层的公寓，能起到屏蔽的作用，朝向东方，能欣赏到城市和博特尼湾的美景。面向主要街道的立面采用随机模式，同时为私人休息空间和室内提供遮荫和隐私保护功能。公寓的造型和内部的细致规划让各个住宅单元的朝向和私密度问题得到完美解决，特别是能俯瞰整个生活区和户外私人空间，以及松木大街上能相互对望的住宅单位之间的空间关系。外部遮阳系统和交错的垂直叶片更进一步地增强了建筑的性能。

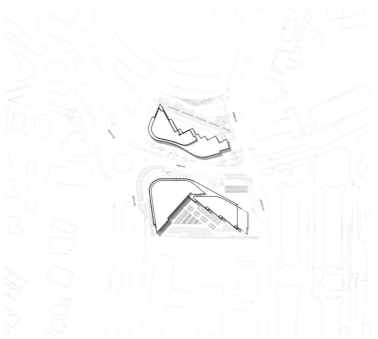

Site Plan
总平面图

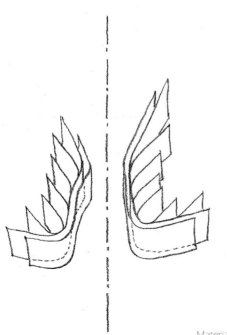

Materiality Principle
原则概念图

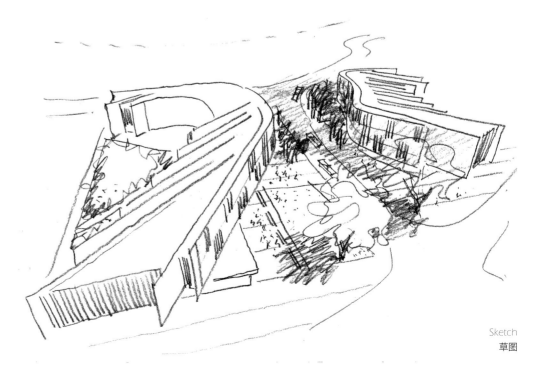

Sketch
草图

In the context of budget and constrained floor area, apartments are designed to maximize efficiency. The majority of plans provide media areas, minimize circulation space and maximize livable space. They are positioned to take advantage of prevailing breezes and promote cross ventilation; the single-oriented apartments are augmented with solar stacks. Each living space aligns with a generous private open space, together with garden terraces and communal landscaped areas, creates amenity for residents as well as attractive foregrounds to views.

Foyers, dispersed along the buildings' primary address, are clearly marked by dramatic awning structures and lightboxes featuring interpretive heritage imagery provided by the adjacent Nursing and Medical Museum. They create a site specific experience for residents and visitors, affirming Prince Henry's past at its most public interface.

在预算和占地面积的限制下，公寓的设计以最大限度提升性能为宗旨，总体规划是为居民提供交流空间，减少流通空间并增加宜居空间。这些空间的布局安排利用了盛行风的优势，促进交叉通风；其中单面朝向的公寓增设了太阳能堆栈系统。每户的生活空间都配有宽敞的私人开放区，包括花园露台和公共景观区，为居民提供舒适的休息场所，近距离地欣赏迷人的风景。

沿着建筑主入口分布的门厅里有显眼的顶棚结构和灯箱，上面印有由相邻护理院和医学博物馆提供的相关文物图像。这些设施为居民和来客提供了独特的体验环境，也是在公共场所宣传亨利王子社区的过去历史。

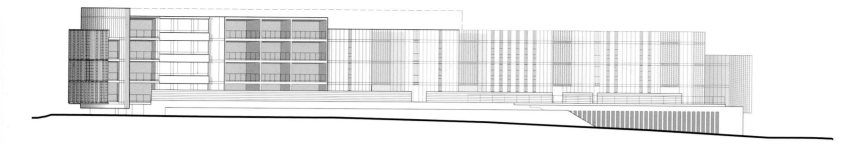

South Elevation
南立面图

West Elevation
西立面图

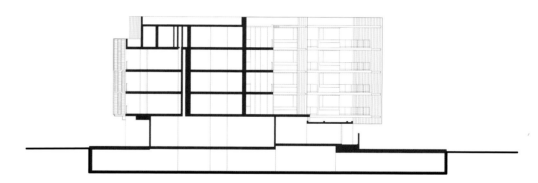

Section1
剖面图 1

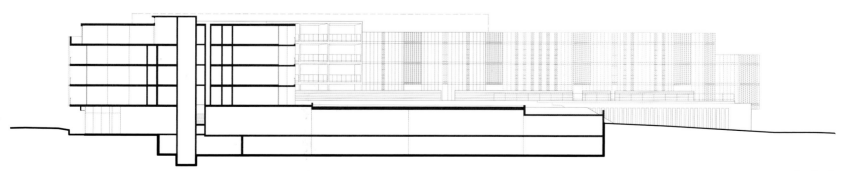

Section2
剖面图 2

SUSTAINABILITY

ESD principles have been integrated into all aspects of design development, construction and operation. PV panels contribute to a renewable energy target of 85.4 tons of CO_2 savings per year for the overall Stockland Development Zone. Solar hot water panels supply approximately 65% of hot water requirements. Grey water recycling provides water for WCs, car washing and irrigation.

可持续性

设计中还遵循了 ESD 原则，融入到了设计开发、建筑和运营的各个方面。光伏电池板提供的可再生能源能让整个斯托克兰开发区每年减少排放 85.4 吨二氧化碳。太阳能热水器能满足 65% 的热水需求。灰水再回收可用来冲洗厕所、洗车和灌溉。

Typical Plan
典型平面图

0   10   20   50m

SHAPE 造型

SUSTAINABILITY 可持续性

FAÇADE 立面

ECOLOGY 生态

303 East 33rd Street, New York, USA

美国纽约东 33 街 303 号

Architect: Perkins Eastman
Client: Toll Brothers, Inc. , The Kibel Companies
Location: Manhattan, New York, USA
Gross Floor Area: 14,307 m²
Floors: 12
Photography: Paúl Rivera/ArchPhoto
Drawings: Perkins Eastman

设计公司：Perkins Eastman
客户：托尔兄弟公司、Kibel 公司
地点：美国纽约市曼哈顿区
建筑面积：14 307 平方米
层数：12
摄影：Paúl Rivera/ArchPhoto
图纸：Perkins Eastman

Developed by Toll Brothers, Inc. and The Kibel Companies, 303 East 33rd Street became the first green development in the Murray Hill neighborhood of Manhattan. The 15,329 m² 12-story building consists of 130 studios, one-, two-, and three-bedroom homes in a variety of layouts. There is also a three-bedroom, four-bathroom triplex penthouse with a total of 158 m² of outdoor space. Amenities include a fully-equipped fitness center, media lounge with pool table, a children's playroom, on-site valet parking, and full-service concierge.

本项目由托尔兄弟公司和 Kibel 公司开发，是曼哈顿 Murray Hill 区的首个绿色项目。这栋面积达 15 329 平方米、高 12 层的大楼共有 130 间工作室，和一室、两室、三室等布局各不相同的公寓。同时还有包含三间卧室、四间浴室的三居顶层公寓，并附有 158 平方米的户外空间。建筑内的设施包括一个设备齐全的健身中心、带有台球桌的多媒体休息室、儿童游戏室，另外还有现场代客泊车服务和全方位的礼宾服务。

MURRAY HILL'S FIRST GREEN CONDOMINIUM
212-777-3303 • 303E33.COM • STUDIO • 1BR • 2BR • 3BR • PENTHOUSE
EXCLUSIVE MARKETING AND SALES

new york sports clubs

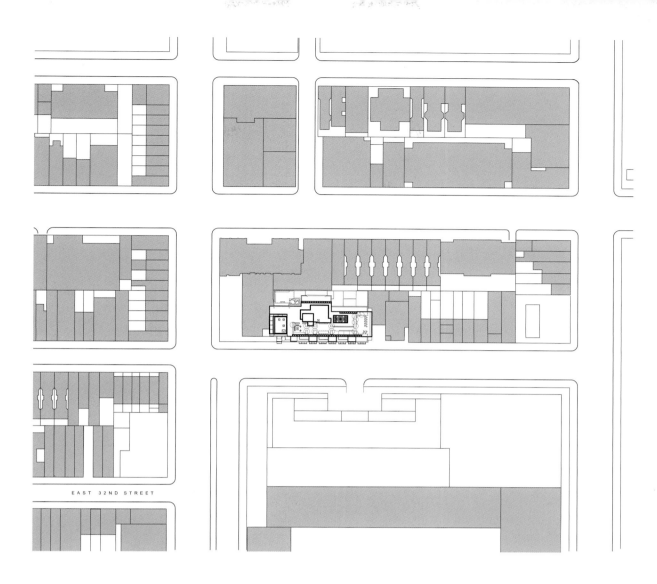

Location Plan
区位图

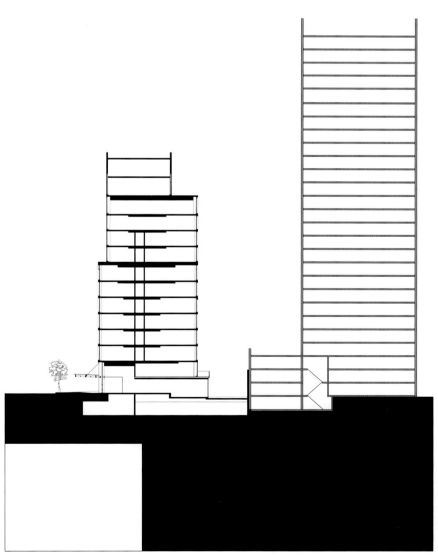

Section
剖面图

SUSTAINABILITY

The project is LEED certified and includes the following sustainable features and NYSERDA incentives for decreased energy demands:

1) Environmental remediation of the site;
2) No new parking was included in the project. Instead, the building offers discounted parking rates for residents with hybrid vehicles in a neighboring parking structure;
3) A waste management plan was created to divert construction waste from landfills;
4) The building envelope is designed to exceed the thermal requirements of NYC codes to reduce both heating and cooling loads. To reduce ozone depletion, all heat pumps use HFC-401A refrigerant;
5) The building will provide electrical sub-metering allowing tenants to monitor their electrical use and manage their energy consumption;
6) The design limits the use of potable water for landscaping by using indigenous plants;
7) The design incorporates low VOC emitting materials and recycled content throughout building.

可持续性

项目获得了 LEED 认证，并具有以下可持续特点，以降低对能源的需求：

1）场地的环境整治；
2）项目并没有建设新的停车场。相反地，建筑为使用混合动力汽车的住户提供贴现的停车费，并引导他们在邻近的停车场里停车；
3）设计师设置了垃圾管理体系，转移堆填区的建筑废物；
4）建筑的立面设计超过了 NYC 代码的要求，以减少冷气和暖气的负荷。建筑还使用了 HFC-401A 制冷剂以减少热泵的臭氧消耗；
5）建筑提供电力分表，让用户可以监测自己的用电量，减少能源消耗；
6）设计多种植本土植物，以减少饮用水的使用；
7）设计在建筑中还采用有机化合物散发量低和可回收的材料。

The building's design is a fresh interpretation of the full- and half-block residential complexes built during the last century, and reflects the mix of architectural diversity in the area. The development is defined as a series of single attached buildings facing the street alternating in height. Brick piers, terraces, balconies, and large expanses of glass fracture the architectural repetition along the street, heightening the concept of an ensemble of buildings rather than a single development.

With quality of life in mind, sustainability is integral to the project's design.

对于建于上世纪的全封闭或半封闭的住宅小区来说，本建筑的设计可算是极富创新性，反映出该区域建筑的多样性。建筑的结构由一系列独立并联系在一起、高度各不相同的大楼组成，面向街道。砖墩、露台、阳台和玻璃的大面积使用将建筑沿着街道的重复立面隔开，加强整体的概念而不是单独的个体。

在提供精神生活的同时，可持续性也是本项目不可或缺的组成部分。

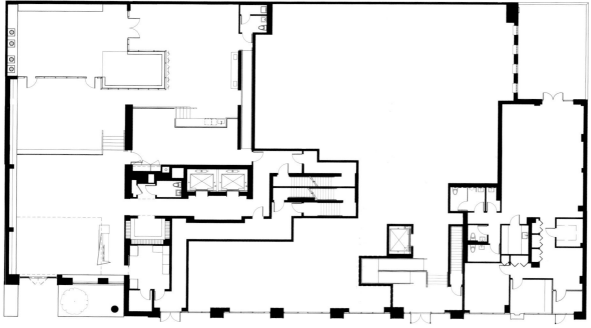

1st Floor Plan
1层平面图

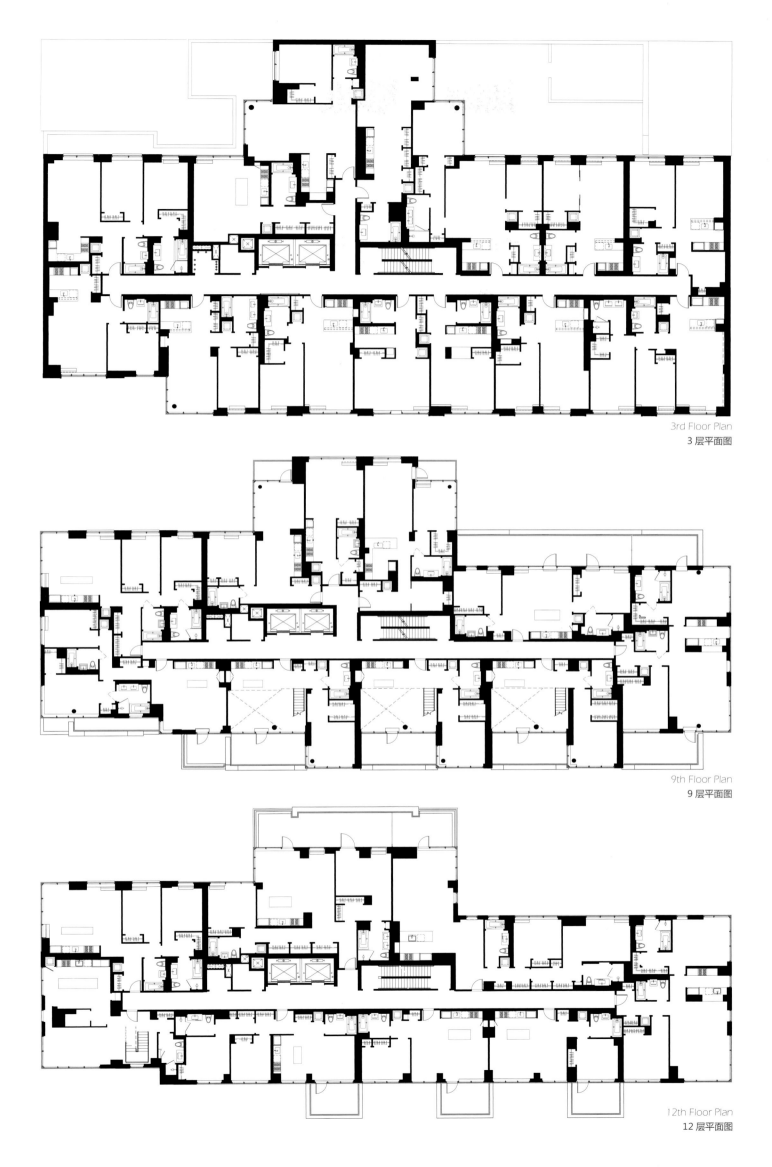

3rd Floor Plan
3 层平面图

9th Floor Plan
9 层平面图

12th Floor Plan
12 层平面图

A landscaped roof-top takes advantage of distinctive urban views of the city and provides an outdoor oasis for the residents. Committed to improve quality of life for the end users, the projects designers create various terraces and balconies, as well as a landscaped roof-top to take advantage of the distinctive urban views of the city that provides outdoor gathering space for the residents' personal use.

屋顶花园坐拥城市美景，为居民提供一处户外绿洲。设计师致力于提高住户的生活质量，设置了各种露台和阳台以及优美的屋顶花园，让住户充分享受到城市景色，或供聚会或供个人使用。

FAÇADE 立面

SHAPE 造型

BALCONY 阳台

ECOLOGY 生态

High Park, Monterrey, Mexico
墨西哥蒙特雷高园综合大厦

Architect: Rojkind Arquitectos
Location: Monterrey, Mexico
Gross Floor Area: 13,000 m²
Floors: 10
Renders: Rojkind Arquitectos, Glessner Group
Drawings: Rojkind Arquitectos

设计公司：Rojkind Arquitectos
地点：墨西哥蒙特雷市
建筑面积：13 000 平方米
层数：10
效果图：Rojkind Arquitectos、Glessner Group
图纸：Rojkind Arquitectos

High Park is located on the outskirts of the northern city of Monterrey, Mexico, surrounded by the Majestic Sierra Madre Oriental Range. The project is designed to take full advantage of its geographic location and to help mitigate the extreme climatic conditions.

本案位于墨西哥北部城市蒙特雷的郊区，四周环绕着宏伟壮观的东马德雷山脉。项目旨在充分利用其地理位置，以缓解极端的气候条件。

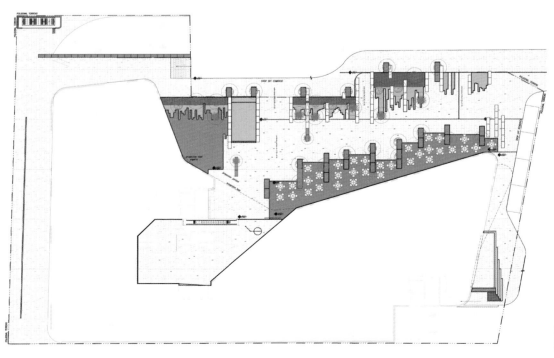

Landscape Plan
景观规划图

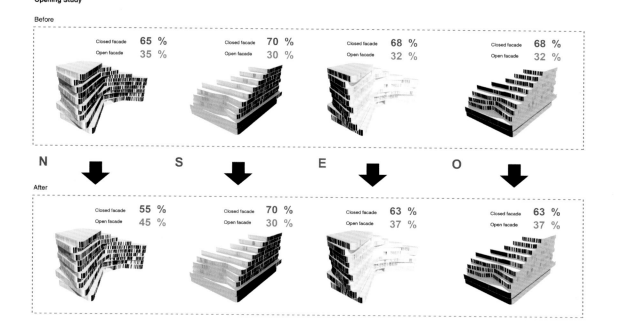

Opening Study

Before

Closed facade **65** %	Closed facade **70** %	Closed facade **68** %	Closed facade **68** %
Open facade **35** %	Open facade **30** %	Open facade **32** %	Open facade **32** %

N   S   E   O

After

Closed facade **55** %	Closed facade **70** %	Closed facade **63** %	Closed facade **63** %
Open facade **45** %	Open facade **30** %	Open facade **37** %	Open facade **37** %

Opening Study
开放度研究

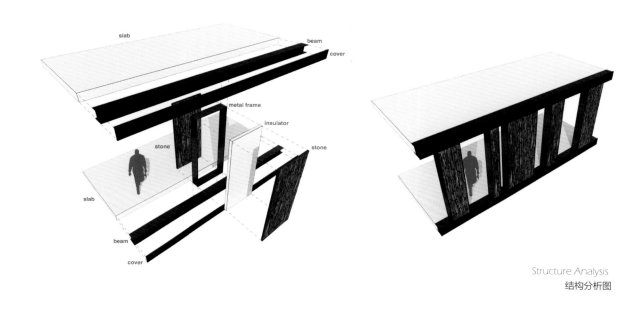

Structure Analysis
结构分析图

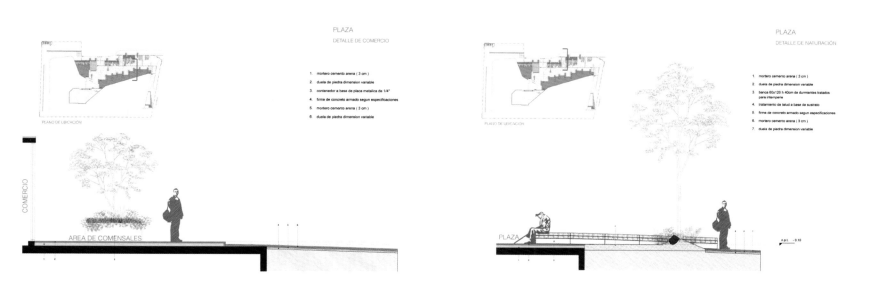

Commercial Plaza Detail
商业广场细节图

To counterbalance the strong sun, the floor plates shift in relation to one another creating a play of light and shadow, and the use of local stone, done by local craftsmen on the façade, allows the building to stay cooler and makes its appearance change as the sun moves across the horizon.

为了阻挡强烈的阳光，各楼层都进行了错位设置，以创造不同的光影效果；另外立面采用由当地工匠打造而成的石材，能让立面保持凉爽，随着太阳的移动外观也会相应地发生变化。

FEATURE 特点分析

SHAPE

As a recurring design concern for Rojkind Arquitectos and as a way of integrating the building into the pedestrian realm (giving back to the community), the building steps back to create an outdoor shaded space that can be enjoyed by the residents and visitors alike.

The project offers outdoor terraces for each apartment due to the strict setback restrictions of the site, capitalizing on the views of the adjacent mountains.

造型

遵循设计师的常规设计理念，建筑要融入社区和行人之中（回馈社会）。
建筑向后缩进，为居民和游客创造一个阴凉的户外空间。

因为场地的缩进限制，建筑能为每间公寓提供室外阳台，饱览周围的山地美景。

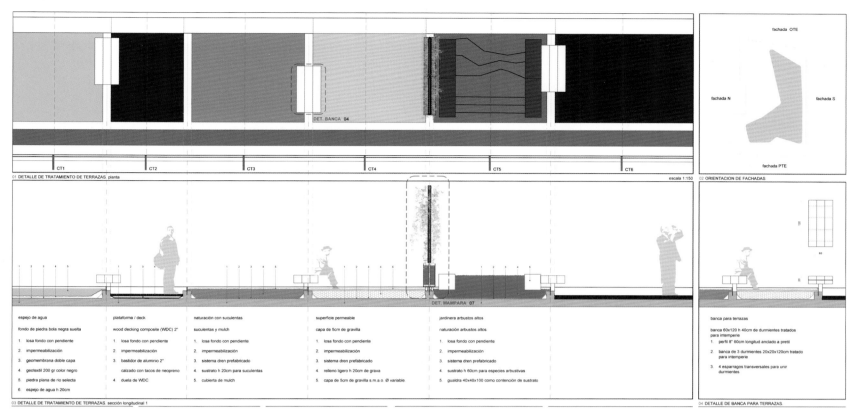

01 DETALLE DE TRATAMIENTO DE TERRAZAS. planta
escala 1:150

02 ORIENTACION DE FACHADAS

fachada OTE
fachada N   fachada S
fachada PTE

DET. BANCA 04
DET. MAMPARA 07

espejo de agua	plataforma / deck	naturación con suculentas	superficie permeable	jardinera arbustos altos	banca para terrazas
fondo de piedra bola negra suelta	wood decking composite (WDC) 2"	suculentas y mulch	capa de 5cm de gravilla	naturación arbustos altos	banca 60x120 h 40cm de durmientes tratados para intemperie
1. losa fondo con pendiente	1. losa fondo con pendiente	1. losa fondo con pendiente	1. losa fondo con pendiente	1. losa fondo con pendiente	1. perfil 8" 60cm longitud anclado a pretil
2. impermeabilización	2. impermeabilización	2. impermeabilización	2. impermeabilización	2. impermeabilización	2. banca de 3 durmientes 20x20x120cm tratado para intemperie
3. geomembrana doble capa	3. bastidor de aluminio 2"	3. sistema dren prefabricado	3. sistema dren prefabricado	3. sistema dren prefabricado	3. 4 esparragos transversales para unir durmientes
4. geotextil 200 gr color negro	calzado con tacos de neopreno	4. sustrato h 20cm para suculentas	4. relleno ligero h 20cm de grava	4. sustrato h 60cm para especies arbustivas	
5. piedra plana de rio selecta	4. duela de WDC	5. cubierta de mulch	5. capa de 5cm de gravilla s.m.a.o. Ø variable	5. gualdra 40x40x100 como contención de sustrato	
6. espejo de agua h 20cm					

03 DETALLE DE TRATAMIENTO DE TERRAZAS. sección longitudinal 1

04 DETALLE DE BANCA PARA TERRAZAS.

Diagram Terraces
露台示意图

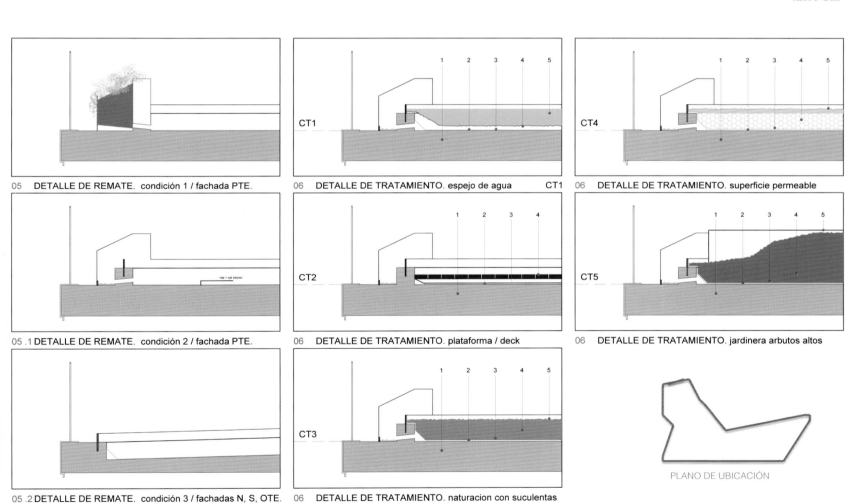

05   DETALLE DE REMATE. condición 1 / fachada PTE.

05 .1 DETALLE DE REMATE. condición 2 / fachada PTE.

05 .2 DETALLE DE REMATE. condición 3 / fachadas N, S, OTE.

CT1
06   DETALLE DE TRATAMIENTO. espejo de agua   CT1

CT2
06   DETALLE DE TRATAMIENTO. plataforma / deck

CT3
06   DETALLE DE TRATAMIENTO. naturacion con suculentas

CT4
06   DETALLE DE TRATAMIENTO. superficie permeable

CT5
06   DETALLE DE TRATAMIENTO. jardinera arbutos altos

PLANO DE UBICACIÓN

ESPEJO DE AGUA	PLATAFORMA/DECK	NATURACION	SUPERFICIE PERMEABLE	ARBUSTOS ALTOS
fondo de piedra bola negra suelta	capa de 5cm de gravilla	wood decking composite (WDC) 2"	naturación arbustos altos	suculentas y mulch
1. losa fondo con pendiente	1. losa fondo con pendiente	1. losa fondo con pendiente	1. losa fondo con pendiente	1. losa fondo con pendiente
2. impermeabilización	2. impermeabilización	2. impermeabilización	2. impermeabilización	2. impermeabilización
3. geomembrana doble capa	3. sistema dren prefabricado	3. bastidor de aluminio 2"	3. sistema dren prefabricado	3. sistema dren prefabricado
4. geotextil 200 gr color negro	4. relleno ligero h 20cm de grava	calzado con tacos de neopreno	4. sustrato h 60cm para especies arbustivas	4. sustrato h 20cm para suculentas
5. piedra plana de rio selecta	5. capa de 5cm de gravilla	4. duela de WDC	5. gualdra 40x40x100 como contención	5. cubierta de mulch
6. espejo de agua h 20cm				

Diagram Terrace Detail
露台细节图

High Park consists of a total of ten levels above ground and three and a half levels of underground parking. The first two levels are for commercial retail, the remaining 8 levels for luxury apartments. Within these 8 residential levels, recreational and entertainment spaces will be provided for the residents including a pool, gym, spa, etc. These 32 apartments will range in size from 250 m^2 to 650 m^2.

Six local designers have been invited to make each apartment unique and appealing to different styles and different market segments. Each apartment has a different layout and configuration, offering a wide range of internal distributions from a one level apartment to a two-story apartment.

项目共包括地上 10 层和 3.5 层的地下停车场。地上底部两层是商店，其余的 8 层是豪华公寓。另有休闲和娱乐空间为居民提供服务，如泳池、健身房、水疗中心等。这 32 间公寓的大小从 250 平方米到 650 平方米不等。

项目还邀请了 6 名本地设计师，将每间公寓都设计得与众不同，自成风格，区域划分也不尽相同。于是每间公寓都拥有不同的布局和配置，无论是单层公寓还是复式公寓，都拥有宽敞的内部空间。

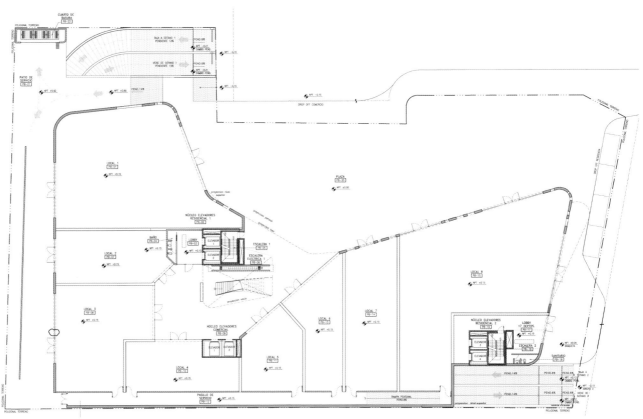

Floor Plan
楼层平面图

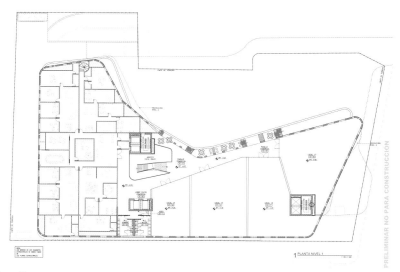

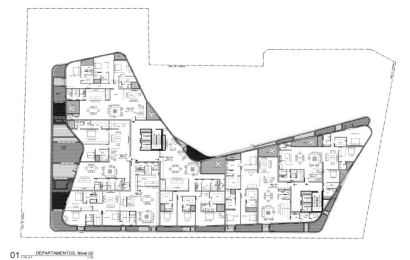

01 ____ DEPARTAMENTOS, Nivel 02

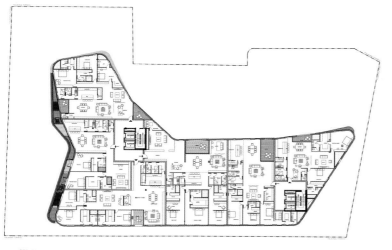

01 DEPARTAMENTOS, Nivel 03

3rd Floor Plan
3 层平面图

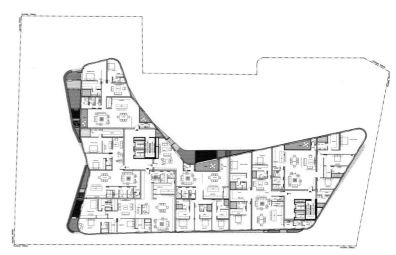

01 DEPARTAMENTOS, Nivel 04

4th Floor Plan
4 层平面图

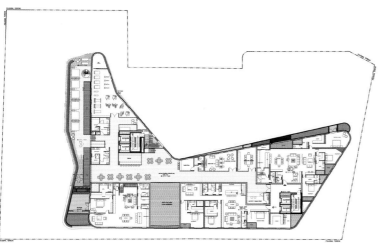

01 DEPARTAMENTOS, Nivel 05

5th Floor Plan
5 层平面图

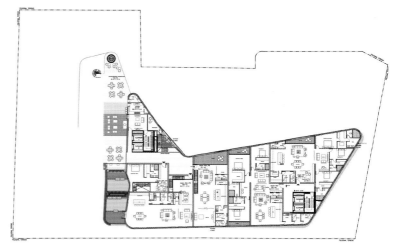

01 DEPARTAMENTOS, Nivel 06

6th Floor Plan
6 层平面图

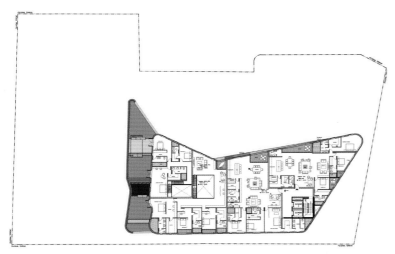

01 DEPARTAMENTOS, Nivel 07

7th Floor Plan
7 层平面图

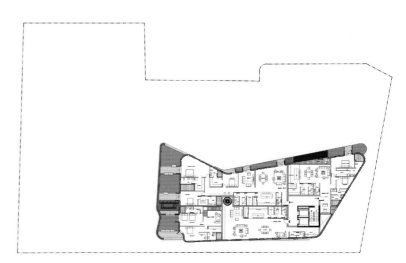

01 DEPARTAMENTOS, Nivel 08

8th Floor Plan
8 层平面图

WINDOW 窗

SHAPE 造型

VIEW 视野

FAÇADE 立面

Iceberg in the Aarhus Docklands, Denmark

丹麦奥尔胡斯港区冰山住宅综合楼

Architect: JDS, CEBRA, Louis Paillard, SeARCH
Location: Aarhus Docklands development, Denmark
Site Area: 25,000 m²
Floors: 10
Photography: Julien De Smedt
Renderings & Model: JDS, CEBRA, LABTOP

设计公司：JDS、CEBRA、Louis Paillard、SeARCH
地点：丹麦奥尔胡斯港区
占地面积：25 000 平方米
层数：10
摄影：Julien De Smedt
效果图 & 模型图：JDS、CEBRA、LABTOP

JDS and Aarhus-based CEBRA, in collaboration with the Dutch firm SeARCH and French architect Louis Paillard won the competition to build a 25,000 m² housing complex in the new Aarhus Docklands Development.

丹麦的 JDS、总部在奥尔胡斯的 CEBRA、荷兰的 SeARCH 公司及法国的建筑师 Louis Paillard 一起在竞争中取胜，共同设计奥尔胡斯港区发展项目中占地 25 000 平方米的住宅综合楼。

The Aarhus Harbour development provides a huge opportunity for Denmark's second largest city to develop in a socially sustainable way by renovating its old, out-of-use container terminal. The area is meant to become a living city quarter, comprised of a multitude of cultural and social activities, a generous amount of workplaces, and of course, a highly mixed and diverse array of housing types.

The Iceberg project seeks to locate itself within the goals of the overall city development. A third of the project's 200 apartments will be set aside as affordable rental housing, aimed at integrating a diverse social profile into the new neighborhood development.

The project's main obstacle is the density set up for the development; the desired square meters are in conflict with the specified site height restrictions and the overall intentions of providing ocean views along with good daylight conditions. The Iceberg negotiates this problematic, by remaining far below the maximum heights at points and emerging above the dotted line at other moments. "Peaks" and "canyons" form, eliciting the project's iconic strength while ensuring that all flats will be supplied with a generous amount of natural lighting and waterfront views.

"With the Iceberg we get unique housing qualities as well as a city architectural expression of the highest quality," says Kent Martinussen, adm. dir. of DAC (Danish Architecture Centre).

"Aarhus will get a fantastic harbour front with unique architectural buildings that both in appearance and functionality prove that we are a city of grand ambitions. Our desire for this area goes beyond just a façade without life and purpose. We want a living city where everybody thrives, both those who live and those who work in this 'City near the harbour / De Bynære Havnearealer'. The projects of this calibre is a big step towards this goal," said Mayor, Nicolai Wammen.

The Iceberg breaks with traditional architecture typically involving a "wall" of homes. The complex takes into consideration those homes located at the back by creating peaks and canyons that ensure optimal light conditions and ocean view.

通过重整其老旧、过时的集装箱码头，奥尔胡斯港区项目为这个丹麦第二大城市的社会可持续发展之路提供了巨大的机遇。港区计划改造成一个充满活力的城市一角，其中包括众多的文化和社会活动空间和开阔的工作场所，当然还有高度混合的一系列多样化住宅类型。

冰山项目组试图在城市总体发展目标中寻找自己的定位。项目总共包括 200 间公寓，而其中的三分之一将被定位为公众可负担得起的出租房，使一个多样化的社会融入到这个新社区的发展中。

项目的主要障碍是开发区规划的密度。所需的面积与指定场地的高度限制以及想拥有海景和良好日照条件的整体意愿相冲突。冰山项目组对此进行了协商，决定最高的地方只保留点，其他地方以虚线连接以形成"峰"和"谷"，突出项目的标志性实力，同时确保所有房间都能享受到充足的自然光和海景。

"多亏了冰山项目组，我们得到了独特的住宅品质，同时还有城市建筑最高品质的表达。"丹麦建筑中心的行政经理 Kent Martinussen 说道。

"奥尔胡斯将会成为一个梦幻的海港，它前面的这栋独特的建筑无论在外观上还是功能上都证明我们是一座富有豪情壮志的城市。我们的愿望不是一个没有生命和目标的建筑，我们想要的是一座朝气蓬勃的城市，使所有在这个港口城市居住或工作的人都感到生机盎然。而拥有如此高品质的这个项目已经朝该目标迈出了一大步。"市长 Nicolai Wammen 说道。

冰山住宅综合楼打破了传统建筑中"墙"的概念。考虑到背面房间的视野，设计师采用"山峰"和"峡谷"式的造型以确保其良好的采光条件和海景视野。

FEATURE 特点分析

SHAPE

The Iceberg was mainly inspired by the amazing location. The architects decided to break up the building into several parts and levels in order to create optimal ocean views for as many apartments as possible. Therefore, the building was split up in a criss-cross manner — inspired by floating icebergs constantly breaking away.

In order to prioritise views and natural light even further, the architects developed sloping roof tops providing stunning views in several directions. The roofs rise and fall in peaks and canyons and, are visually displaced from each other just like the motions of real icebergs. Creating a logical harmony with the heaven-aspiring triangular spires, the façade was split into loosely combined triangles that lend the façade a dynamic feel. This provides a harmonious connection between the complex as a whole and its details. The large white triangles and the flickering surface gave rise to the name of the complex: The Iceberg.

造型

冰山的灵感来源于其令人惊叹的地理位置。设计师决定把建筑分为几个部分和层次，努力让最多的房间欣赏到美丽的海景。因此，受漂浮的冰山经常分裂这种现象所启发，建筑被分裂成纵横交错的形式。

为了使视野更宽广，自然采光更充足，设计师采用了倾斜的屋顶，以保证不同方向的开阔视野。屋顶的上升和下降形成了"山峰"和"峡谷"，看起来就像真正的冰山一样。高耸入云的三角尖顶一座连着一座，把立面变成了松散结合的三角形，看起来更具动感，把建筑的整体和细节都和谐地融合在一起。而这些大型的白色三角形和闪耀的立面就形成了建筑的名字——冰山住宅综合楼。

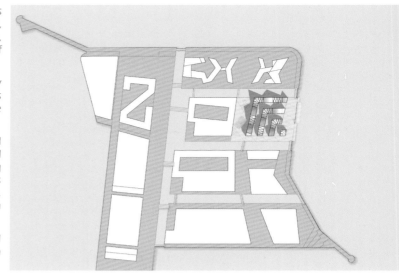

Location Plan
区位图

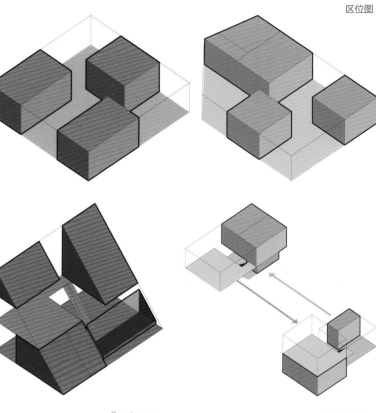

Penthouse
复式套房

Urban Villa
城市别墅

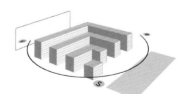

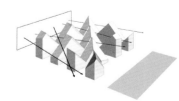

Sun & Views Analysis
光照 & 视野分析图

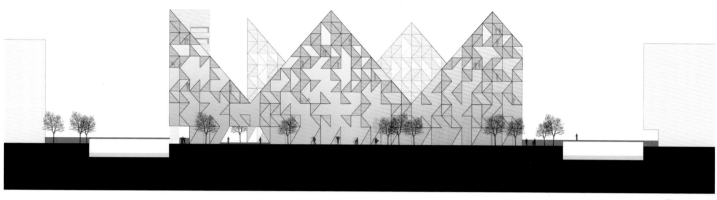

Elevation 1
立面图 1

Elevation2
立面图 2

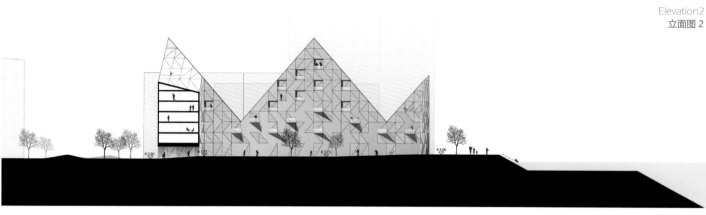

Section1
剖面图 1

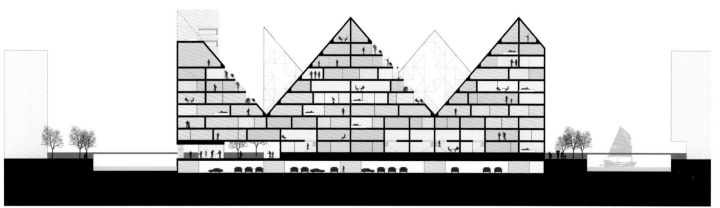

Section2
剖面图 2

Section3
剖面图 3

Ground Floor Plan
首层平面图

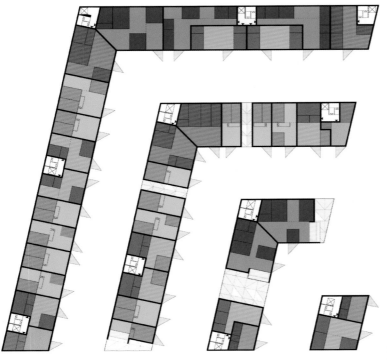

2nd Floor Plan
2 层平面图

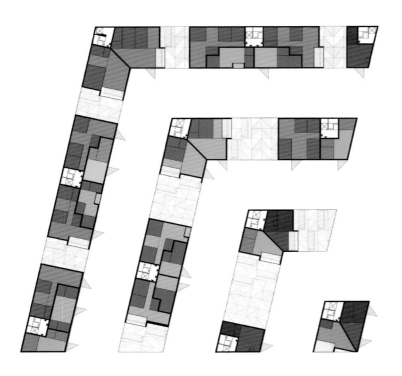

4th Floor Plan
4 层平面图

7th Floor Plan
7 层平面图

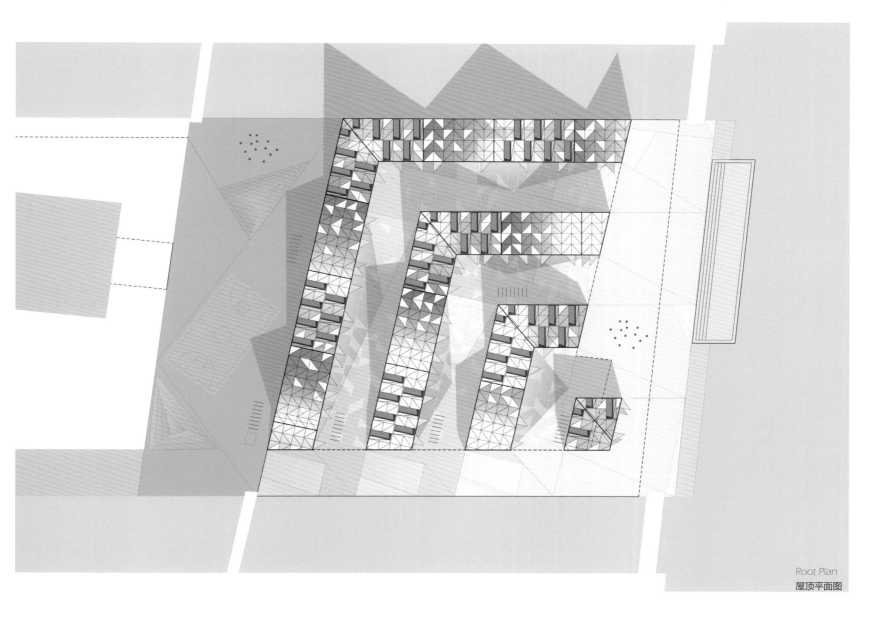

Root Plan
屋顶平面图

SUSTAINABILITY 可持续性

SHAPE 造型

STRUCTURE 结构

OPEN SPACE 开放空间

Green School Stockholm, Sweden

瑞典斯德哥尔摩绿色学校

Architect: 3XN
Location: Stockholm, Sweden
Site Area: 22,800 m² + 1,600 m² (greenhouse)
Floors: 14

设计公司：3XN
地点：瑞典斯德哥尔摩市
占地面积：22 800 平方米 + 1 600 平方米（温室）
层数：14

Green School Stockholm is a new type of school with a modern approach for sustainable living. By actively educating about locally grown food, and by creating a multitude of green exterior public spaces, the building encompasses a full lifetime of sustainable living: from kindergarten to high school, college dorms to senior apartments.

本案这是一座新式的学校，表达了现代的可持续生活理念。学校积极教育学生种植本地植物，从而打造出大量的绿色户外空间。学校还有完整的可持续生活体系，能供应从幼儿园到高中到大学的宿舍，甚至是老年公寓的需求。

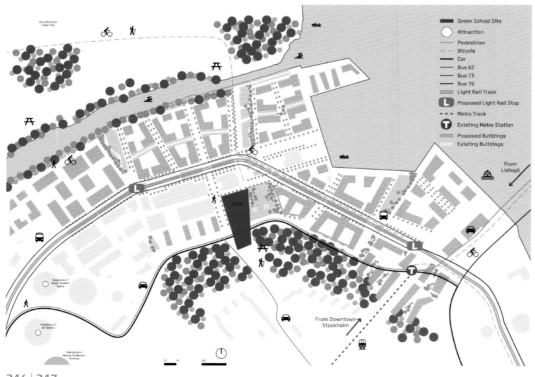

The building is formed by two adjoining arcs. The green school and accompanying greenhouse constitute the public arc and allow for internal and external circulation through the building with the vegetation growing all around. Nine levels of housing for students and seniors twist, slide and shift to create wide private terraces and maximum daylight exposure.

建筑由两个相连的拱形组成。绿色学校及温室组成了公共的拱形，它使内外的流线和绿色的环境时刻联系在一起。学生住宿和老年公寓的九层空间相互交织、转换，形成宽阔的私人阳台，有助于采光。

Location Plan
区位图

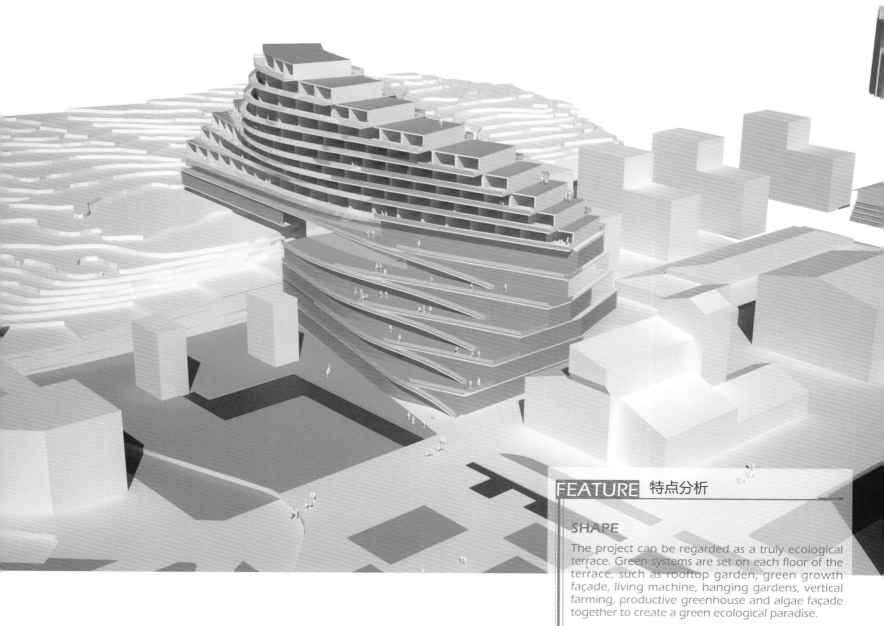

SHAPE

The project can be regarded as a truly ecological terrace. Green systems are set on each floor of the terrace, such as rooftop garden, green growth façade, living machine, hanging gardens, vertical farming, productive greenhouse and algae façade together to create a green ecological paradise.

造型

本项目可称得上是名符其实的生态梯田。建筑梯田的造型上每层都留出大量空间种植绿色植物，如屋顶花园、绿色外墙、生态系统、悬浮花园、垂直农业、高产温室、藻类立面等，共同打造出绿色的生态天堂。

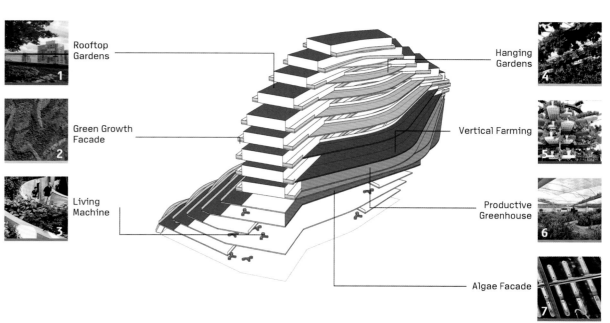

Rooftop Gardens — 1

Green Growth Facade — 2

Living Machine — 3

Hanging Gardens — 4

Vertical Farming — 5

Productive Greenhouse — 6

Algae Facade — 7

Plants Plan
植被规划图

Wide atriums open up the green school to accommodate spontaneous learning. This green pathway through the school culminates with a large greenhouse as the focal point. The greenhouse encompasses three enclosed levels for maximum productive growing and extends upwards with hanging gardens and vertical farming alongside the student and senior residences. Inside the public area of the building is an organic food store, where the organically grown vegetables from the greenhouse are sold. The Green School's kindergarten is located directly adjacent to a birch grove. Here, the children get their own safe oasis.

Green terraces outside the building allow pedestrians to ascend the building and move from the lower northern side of the site up and across the street to the higher southern side. The building thus becomes a productive extension of the planned green corridor for the area as well as an avenue for the public.

开阔的中庭则作为绿色校园里学生自发学习的绝好场所。绿色的通道的最高处是一个巨大的温室,温室包含3层封闭的楼层,以获取最佳种植环境。通道和漂浮花园一同向上延伸,另外沿着学生宿舍和老年公寓的还有垂直农业等元素。

行人可以通过外部的大型绿色梯田走进建筑,从较低的北侧穿过道路一直来到较高的南侧。本项目因此成为本区域的绿色走廊计划的高产扩展部分,也作为公共聚会的场所。

BALCONY 阳台

SHAPE 造型

FAÇADE 立面

MATERIAL 材料

Fiera Milano, Milan, Italy

意大利米兰国际展览中心综合项目

Architect: Studio Daniel Libeskind
Client: CityLife
Location: Milan, Italy
Site Area: 288,878.9 m²
Renderings: Studio Daniel Libeskind, Hayes Davidson, Stack Studio, Studio AMD
Models: Studio Daniel Libeskind
Drawings & Plans : Studio Daniel Libeskind

设计公司：丹尼尔·里伯斯金工作室
客户：城市生活发展公司
地点：意大利米兰市
占地面积：288 878.9 平方米
效果图：丹尼尔·里伯斯金工作室、海耶斯·戴维森、斯塔克工作室、AMD 工作室
模型图：丹尼尔·里伯斯金工作室
图纸 & 规划图：丹尼尔·里伯斯金工作室

The Fiera Milano project is located in the heart of Milan, spanning 428,966.78 m² on the old fairgrounds of this historic city. The scheme of Fiera Milano incorporates residential and office development, retail space and a museum all built around a central park, a much needed open space within the city. In 2004, Studio Daniel Libeskind along with Zaha Hadid Architects, and Arata Isozaki & Associates won the commission for the development of the site. In addition to leading the master plan, SDL is designing the contemporary art museum, the central office tower, a residential tower, and the first housing parcel.

米兰国际展览中心综合项目位于米兰的中心地带，占地约 428 966.78 平方米，这里曾经是这座历史古城的旧会展中心。项目包括住宅、办公、零售商店和博物馆，全都围绕着中央公园。这里将满足城市需求的开放空间。在 2004 年，丹尼尔·里伯斯金工作室、扎哈·哈迪德建筑师事务所和矶崎新事务所一起合作赢得了场地的开发委托项目。除了总规划外，丹尼尔·里伯斯金工作室还设计了一座当代艺术博物馆、中央办公楼、住宅楼，并首次进行住宅的外观设计。

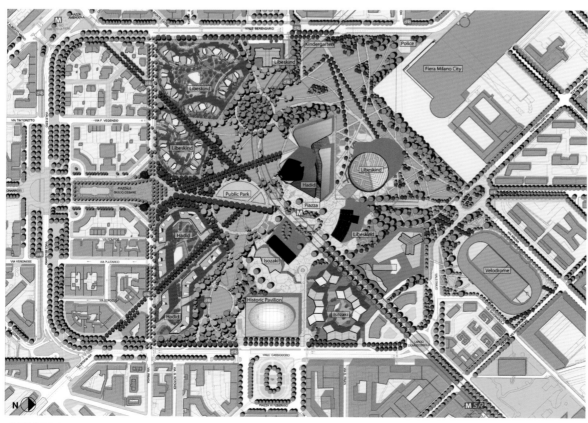

Site Plan
总平面图

The project is large, in both scope and area, and will ultimately create a new neighborhood in the cultural and historical context of Milan. The challenge posed by the Fiera site, which is central and critical to the development of Milan, is that it should not be viewed as merely a building development, but something more all together. Milan is a cultural center for Italy, exhibiting the best of what the country has to offer. It is a place that holds the dreams, aspirations and pride of all the Milanesi. It is in this way that the site must be representative of the greatness of Italian design, furniture, fashion technology and deserves nothing less than a visionary, yet practical, urban scheme.

本项目无论规模和范围都很巨大，建成后将会成为具有文化和历史内涵的米兰新社区。项目场地带来的挑战对米兰的日后发展极为关键，即项目不应该只是发展建筑，而该加入更多的东西。米兰是意大利的文化中心，向世界展示着这个国家里最好的一切。米兰是一个拥有梦想、激情和荣誉的地方。正因为如此，项目必须是一个长远、实用的城市规划，成为可以装载意大利设计、家居、时尚、科技的场所。

FEATURE 特点分析

SHAPE

The design breaks the mode of rigid lines as the asymmetric balcony creating a curved line attached the outside façade. The façade covered by glass and slabs not only ensures vision, but provides privacy protection.

造型

建筑设计突破了呆板的直线条，阳台呈不对称设计，营造出的曲线形线条附着在外立面上。同时立面玻璃和条形板的混合搭配，不仅保证了视野，也为住户提供了隐私保障。

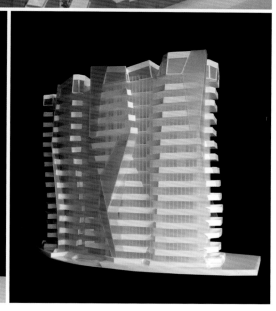

Piano 14 +52.50 m
Piano 13 +48.90 m
Piano 12 +45.30 m
Piano 11 +41.70 m
Piano 10 +38.10 m
Piano 9 +34.50 m
Piano 8 +30.90 m
Piano 7 +27.30 m
Piano 6 +23.70 m
Piano 5 +20.10 m
Piano 4 +16.50 m
Piano 3 +12.90 m
Piano 2 +9.30 m
Piano 1 +5.70 m
Piano Terra +1.50 m

Section
剖面图

ROOFTOP GARDEN 屋顶花园

SHAPE 造型

FAÇADE 立面

STRUCTURE 结构

RATP Apartment, Paris, France

法国巴黎 RATP 公寓

Architect: ECDM
Client: RATP
Location: Paris, France
Site Area: 6,560 m²
Floors: 7
Completion: 2013

设计公司：ECDM
客户：RATP
地点：法国巴黎市
占地面积：6 560 平方米
层数：7
完成时间：2013 年

Situated in Paris, near the Porte d'Orléans, this project partakes in the extensive restructuring of an RATP industrial site spanning 17,000 m², initiated with the aim of developing and modernizing the existing bus center. Rue du Père Corentin, tight and narrow, like the surrounding district, is characterized by working class housing, real estate developments indifferent to the context and a few apartment blocks. The access to the bus center from Rue du Père Corentin being invariable, the project is developed above the access ramp for buses and opens onto the rooftop of the bus center which is converted into a huge suspended garden.

本案坐落在巴黎，靠近 Porte d'Orléans 酒店，是占地 17 000 平方米的 RATP 工业区改组扩建的一部分，目的是同时发展原来的巴士中心，使其更为现代化。和周围的区域一样，Rue du Père Corentin 区地形狭小，最大的特点就是聚集了很多工人住房和少数公寓楼，很多开发商都对这里的环境持冷淡态度。从 Rue du Père Corentin 区通向巴士中心的道路要保持原状，因此项目只能在巴士通道的上方建造，并连接到巴士中心的屋顶，将屋顶转换成一座巨大的空中花园。

SHAPE

The two buildings of the project echo with each other in a certain angle that will hold enough private spaces for residents. Trapezoidal structure is the highlight of the project. The balconies retreat as height rises, which will reduce the volume of the building. This is to reduce the impact on adjacent buildings and surrounding environment; and this innovative design will bring vitality to the city. As a result, the project appears like a white galleon or modern pyramid.

造型

建筑的两栋体量成一定角度组合，相辅相成又为彼此留出足够的私密空间。体量的梯形结构是建筑的最大特色，层层向上缩进的阳台将建筑体积逐渐缩小，目的是减少对邻近建筑和周围环境的影响，同时以新颖的创意为城市带来新的活力。最终呈现出的建筑造型就像一艘白色帆船，又似一座现代金字塔。

The project consists of two buildings that echo with each other. The specific work on this project takes as a reference and support such emblematic constructions in Paris as Eugène Beaudoin and Marcel Lods building Passage d'Atlas, or more precisely still Henri Sauvage's building Rue Vavin. Stepped terraces rise from the street and turn at angles to the right of the two contiguous houses, and continue above the suspended garden. The whole is founded not on nostalgia or formalism, but on the desire to open up to the street and open the sky to passersby.

The two façades do not so much face each other as they complement and support one another in an interplay of cascades that simultaneously releases and captures light. An oblique suspended garden appears, fragmenting the presence of greenery in a multitude of planes. Like some white galleon, with a prow in polished concrete, we have here an expression of Parisian urban regulations pushed to their highest pitch. The white lacquered envelope concentrates and diffracts the ambiances of the suspended gardens, forming a surprising contrast to the extreme mineral quality of the street.

The result is a modern pyramid lined with greenery, a monolith whose paradoxical lightness stands in contrast to the thickness of the surrounding buildings.

The project is thus hyper-contextual, concretizing a response shaped by the constraints of the site. The positioning between two landscapes and two urban templates suggests a morphing between very two very characteristic urban elements, two periods and two modes of living.

本案由两栋大楼组成，相互呼应。建筑采用的建造技术参考了建筑师尤金·波多因和马塞尔·罗兹建造的 d'Atlas 通道，更确切地说还有亨利·索维奇的 Rue Vavin 建筑。阶梯式的阳台从街面开始上升，在两栋大楼连接处向右转角，连接到最顶层的屋顶花园。建筑整体既不传统又不形式主义，又满足向街道开放、将天空还给人们的愿景。

两栋大楼的立面并不完全对立，而是以阶梯的形式相辅相成，同时适量释放和捕捉光线。倾斜的屋顶花园将绿色分洒到各个角落。建筑就像是一艘船头由抛光混凝土打造的白色帆船，设计师希望借此表达出巴黎城市法规的最高点。白色漆面吸收并衍射光线到屋顶花园，和矿质街道形成强烈的对比。

最终建成的建筑是一座现代的金字塔，两旁绿意葱葱，如同一块轻质的巨石矗立中间，周围被厚实的建筑围绕，对比鲜明。

因此，本案是一座超越环境的建筑、将场地限制转化成具体的应对策略。位于两个景观和两个城市模板之间，这就要求了建筑必须进行变形，形成两座特色的城市元素，容纳了两个时期和两种生活模式。

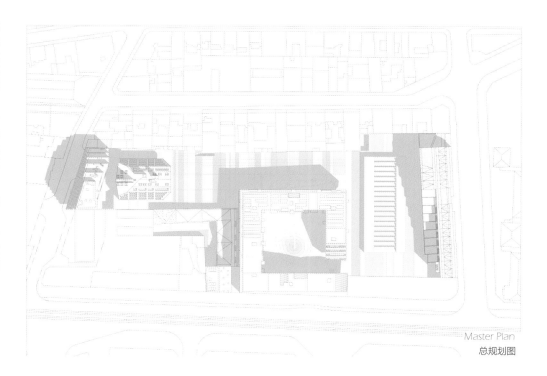

Master Plan
总规划图

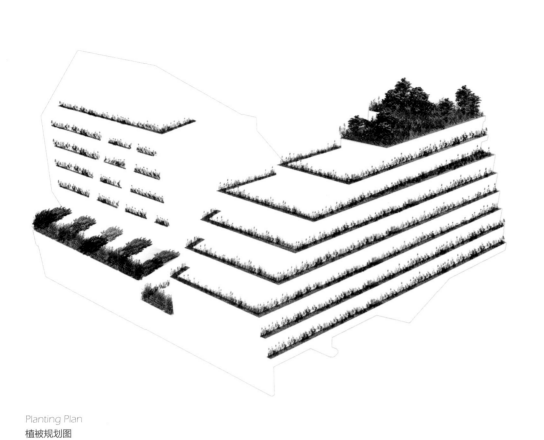

Planting Plan
植被规划图

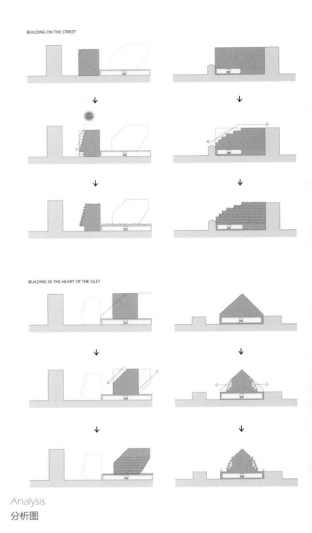

BUILDING ON THE STREET

BUILDING IN THE HEART OF THE ISLET

Analysis
分析图

BATIMENT A

N+1   N+2   N+3   N+4   N+5   N+6   N+7

Tower A Structure
A 楼结构图

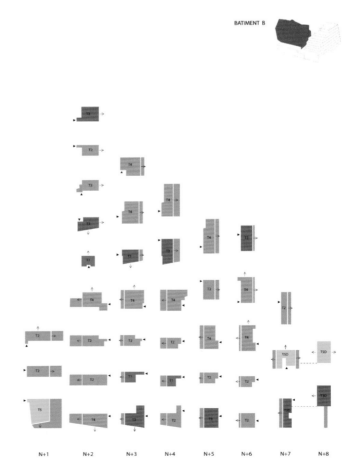

BATIMENT B

N+1   N+2   N+3   N+4   N+5   N+6   N+7   N+8

Tower B Structure
B 楼结构图

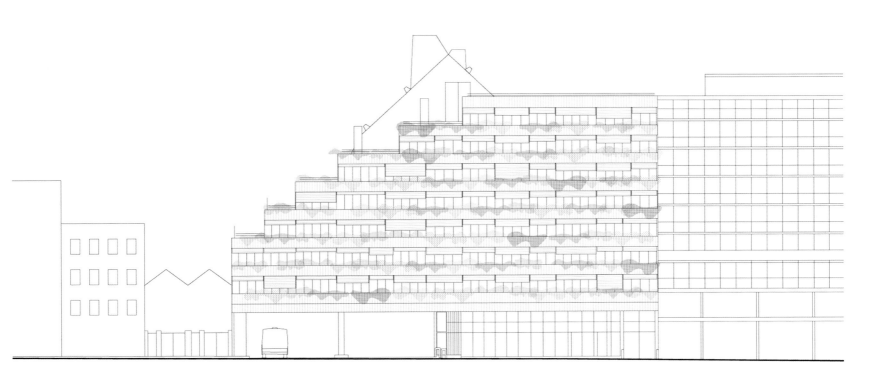

Elevation

立面图

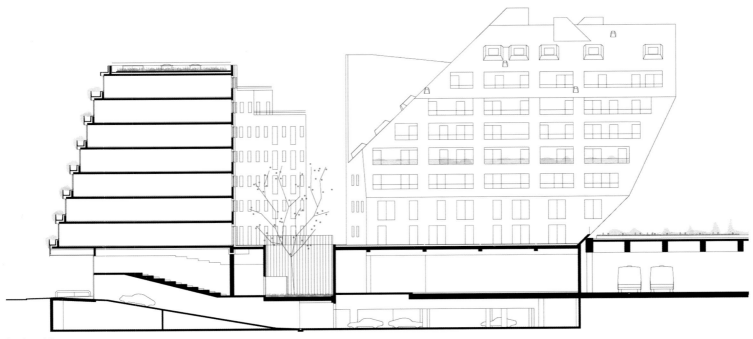

Section AA
剖面图 AA

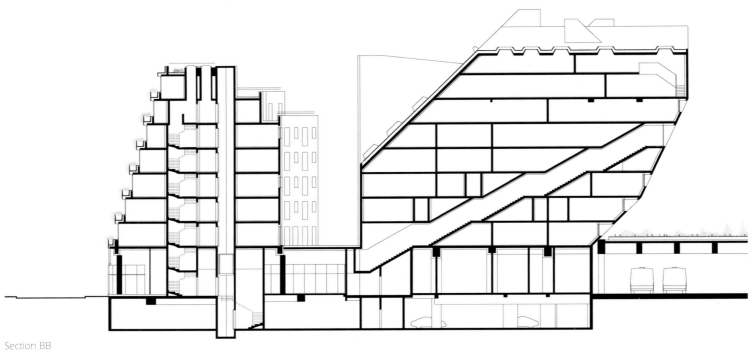

Section BB
剖面图 BB

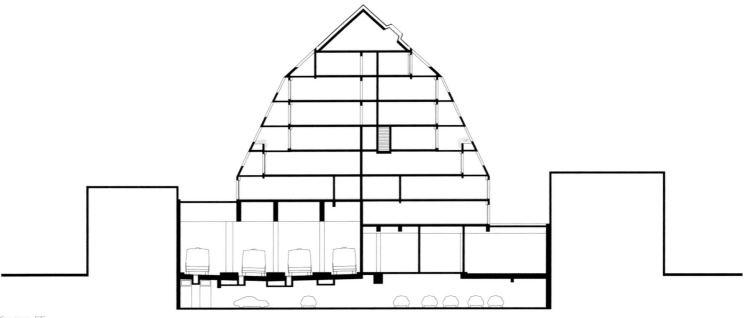

Section EE
剖面图 EE

In the framework of a reflection on the city, the subject lends itself to a mix, associating a bus depot and housing in a way that harnesses the very 19th-century character of the industrial terrain to create a contemporary living environment.

In an attempt to embed the project into this heteroclite district with an unstructured cityscape, it is organized around two converging intentions:

1) To restore a sense of urban continuity by reestablishing a relationship between the foundations of the building. With sections of façades cut on the bias and effects of folding, the project is a proposition in framing boundaries that emphasizes the fluidity of the public sphere and attempts to offset the abrupt setback of the adjoining building.

2) To maximize the views, light, privacy, space, and a sense of openness to the city, systematically on all levels, as much for the residents of this project as for those in neighboring projects and for people in the public space.

在反映城市框架的基础上，建筑将自身和巴士仓库、住房相结合，运用 19 世纪工业区的地形特点营造出当代的生活环境。

设计师尝试着将项目融入此不规则的街区和零散的城市景观中，围绕着两个目标进行：

1）恢复城市连续性，重筑建筑与城市的关系。通过斜切的立面和折叠效果，项目实际上是定义出场地的边缘，强调公共领域的流动性，并尝试弥补对毗邻建筑物产生的冲突。

2）最大限度保证视野、采光、私密性、空间利用和对城市的开放感，各个楼层都井然有序，为本案公寓的居民、邻近大楼及在公共空间活动的人们提供方便。

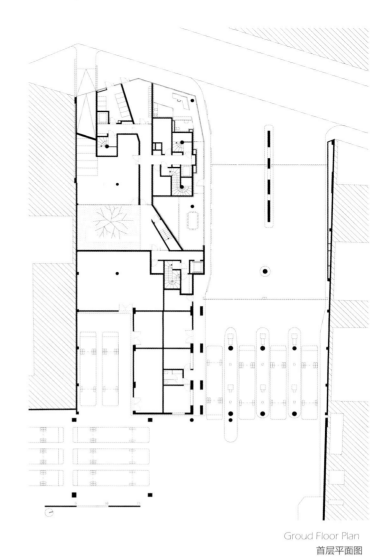

Groud Floor Plan
首层平面图

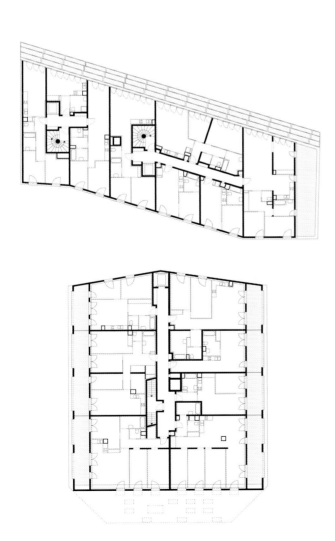

3rd Floor Plan
3 层平面图

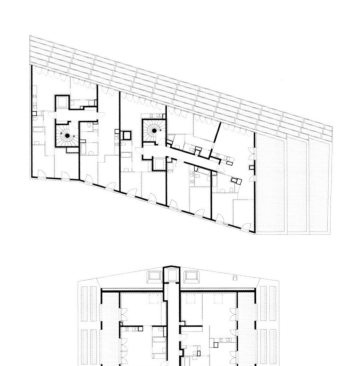

5th Floor Plan
5 层平面图

FAÇADE 立面

SHAPE 造型

STRUCTURE 结构

WINDOW 窗

NE Apartment, Tokyo, Japan

日本东京 NE 公寓

Architect: Nakae Architects, Akiyoshi Takagi Architects, Ohno JAPAN
Location: Tokyo, Japan
Site Area: 201.89 m²
Gross Floor Area: 289.02 m²
Floors: 3
Height: 8.05 m

设计公司：Nakae Architects、高木昭良建筑设计事务所、Ohno JAPAN
地点：日本东京市
占地面积：201.89 平方米
建筑面积：289.02 平方米
层数：3
高度：8.05 米

This 8-unit rental apartment house complex was designed to house motorcycle enthusiasts, with a built-in garage included in every unit.

The building is located on a flag-shaped plot near the apex of a triangular block, with a certain degree of open space toward the main road to the south.

这套包含 8 套出租公寓的住宅是专为摩托车爱好者而设计的，每间房内都有一间内置车库。

建筑坐落的场地呈旗形，位于一片三角形街区的顶点，朝南面向主要道路，有一定的开放空间。

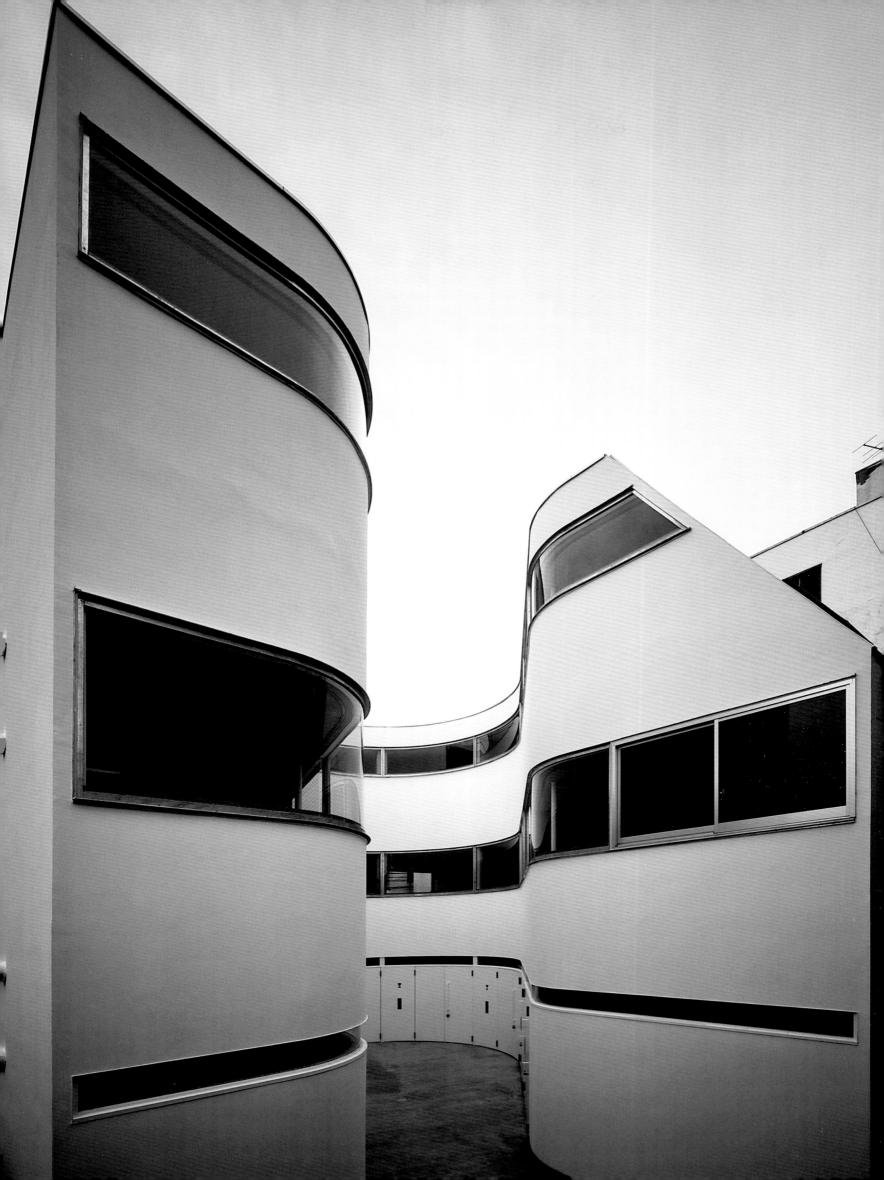

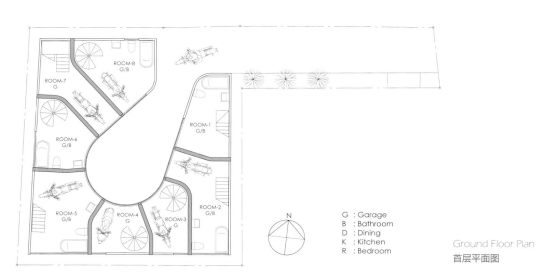

G : Garage
B : Bathroom
D : Dining
K : Kitchen
R : Bedroom

Ground Floor Plan
首层平面图

Location Plan
区位示意图

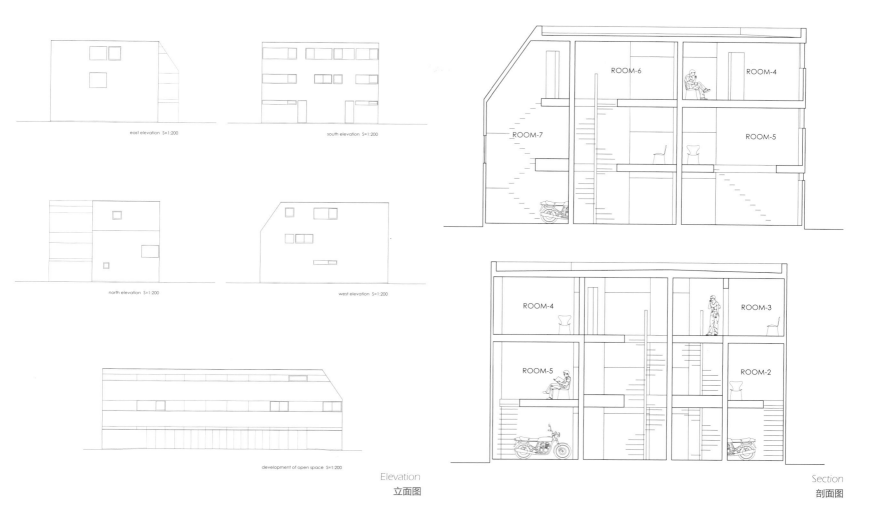

east elevation S=1:200

south elevation S=1:200

north elevation S=1:200

west elevation S=1:200

development of open space S=1:200

Elevation
立面图

Section
剖面图

FEATURE 特点分析

SHAPE

The C-shaped design was a practical decision to allow the residents to access their apartments through a common alley that leads right to the center of the complex. The wall on the entrance side was curved to provide maximum space on the outside, while guaranteeing sufficient volume for each apartment unit and wall length to fit 8 entrance doors. The resulting little square avoids giving the impression of a narrow and dark dead end, and allows the residents to rotate their bikes easily.

造型

建筑的 C 形造型极具实用性，为居民提供了一条位于公寓的正中心的通道，明确地引领居民进入公寓。入口处的墙壁呈弯曲状，为外部提供最大的空间，同时保证每间公寓能有足够的空间，墙壁的长度也要适合 8 间公寓的门口。由此而成的小广场能避免造成狭窄和黑暗的死角，并且方便居民的车子调头。

The building is a reinforced concrete structure composed of seven walls and a slab. The main characteristic of the structure lies in the fact that the reinforced walls, composed of an in-plane rigid frame of columns and beams, were disposed in a radial pattern. The walls rely on the transfer of horizontal force from the slabs instead of using perpendicular beams. They are in fact vertical cantilevers fixed in the foundation of the building. Although the centrally-oriented radial displacement is vulnerable to rotational forces, the changing angles of each wall reinforce the structure's resistance.

Because the structure of the building relies on the seven interior walls, the exterior wall was handled to use a dry construction method. This allowed us to continue studying the emplacement and size of the wall openings in accordance with the uneven surroundings until the very last moment of the construction process. The functions of each wall are also enhanced by a clear division of their roles: structure and sound insulation for the interior walls, openings and thermal insulation for the exterior walls. Despite their curve, the interior walls always meet the outside wall at right angles, preventing the presence of sharp corners and thus improving livability.

On the entrance side, each floor is fitted with a continuous strip of curved windows, with a comparatively wider opening on the second level. The orientation of each room was set to avoid a direct view of the opposite apartment. Combined with a double-panned window, this setting provides a peculiar feeling of privacy.

建筑采用钢筋混凝土结构，由 7 面墙壁和一块楼板组成。结构的主要特征在于其钢筋墙是由立柱和横梁组成的平面刚性框架，呈放射状分布。墙壁不是采用垂直梁，而是依靠楼板来转移水平压力。它们其实是安装在建筑地基上的垂直悬臂，虽然放射布局的中心容易受到旋转力的影响，但角度都有所不同的每面墙壁能加强了结构的抵抗力。

由于建筑的结构主要依赖 7 面内墙，外墙就使用了干法施工。这也方便设计师深入研究墙壁开口的位置和大小，以便在施工时能随时调整以适应弯曲的造型。每面墙的功能也得到了提升，且分工明确：内墙用于支撑结构和隔音，外墙则设置开口采光和隔热。尽管墙壁是弯曲的，但内墙总能和外墙形成直角，避免产生锐角，提升宜居性。

在入口处一侧，每层楼都设施了和弯曲墙壁相适应的连续长形窗口，在二层则是相对较宽的窗口。每间房间的朝向都刻意避免对面房间的直视，结合双窗格的窗口，更加强各户的私密感。

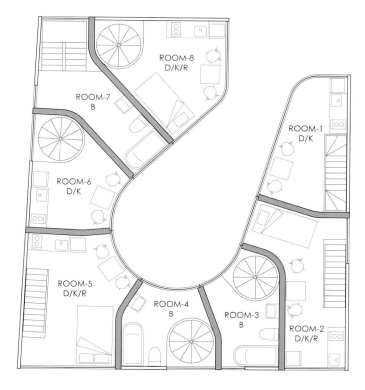

ROOM-8
D/K/R

ROOM-7
B

ROOM-1
D/K

ROOM-6
D/K

ROOM-5
D/K/R

ROOM-4
B

ROOM-3
B

ROOM-2
D/K/R

2nd Floor Plan
2 层平面图

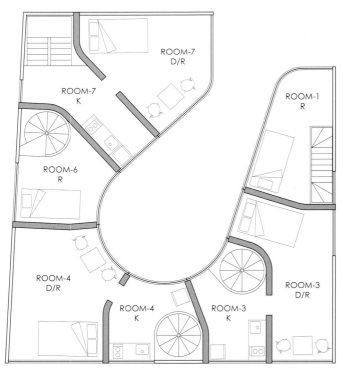

The walls separating each apartment unit were disposed in a radial pattern, each with a gentle curve that leads them to meet the external wall at a right angle. By connecting the angles of each room, the curved walls contribute to give the impression of a more spacious environment. The entire structure is designed as an extension of the road, smoothly following the movement of its residents as they drive through the alley, enter the central square, park their motorcycle in the garage and move upstairs to their living quarters.

墙壁将每个房间分割成放射状布局，每间房都有轻微的弧度来适应外墙的角度。通过每个房间的角度衔接，能让公寓获得更宽敞的空间。整个结构都设计成道路的延伸，顺应居民的行动路线：开车通过入口通道，进入中央广场，停车到车库然后上楼回到自己的房间。

3rd Floor Plan
三层平面图

STRUCTURE 结构

SHAPE 造型

ECOLOGY 生态

FAÇADE 立面

Beirut Marina & Town Quay, Beirut, Lebanon

黎巴嫩贝鲁特码头和城镇港口综合体

Architect: Steven Holl Architects, Nabil Gholam Architecture
Client: Solidere
Location: Beirut, Lebanon
Gross Floor Area: 20,439 m²
Floors: 5

设计公司：Steven Holl Architects、Nabil Gholam Architecture
客户：Solidere
地点：黎巴嫩贝鲁特市
建筑面积：20 439 平方米
层数：5

The building is situated along a new fabricated terrain, which is extended from Beirut's Corniche, the seaside promenade, to create an "urban beach" of public spaces overlooking the marina.

建筑坐落在新规划建设的城区中，从贝鲁特滨海路一直延伸到滨海长廊，为公共打造了一处"城市沙滩"，可俯瞰海滨美景。

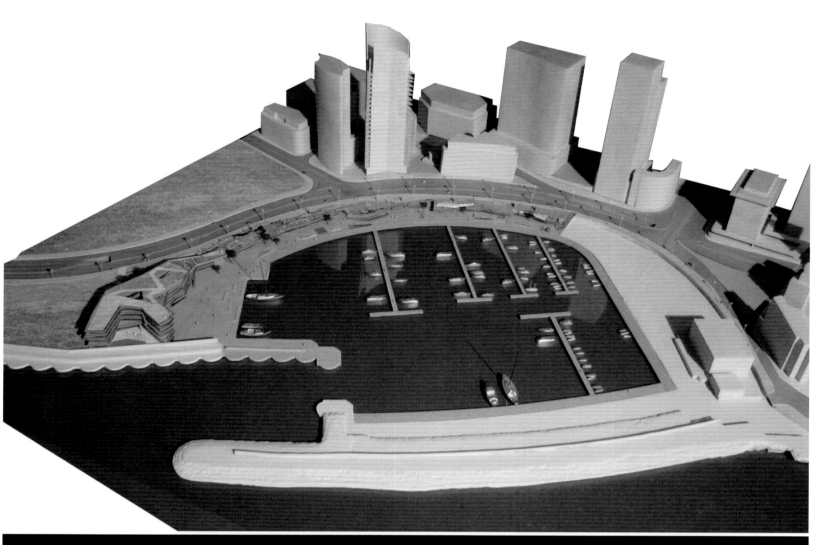

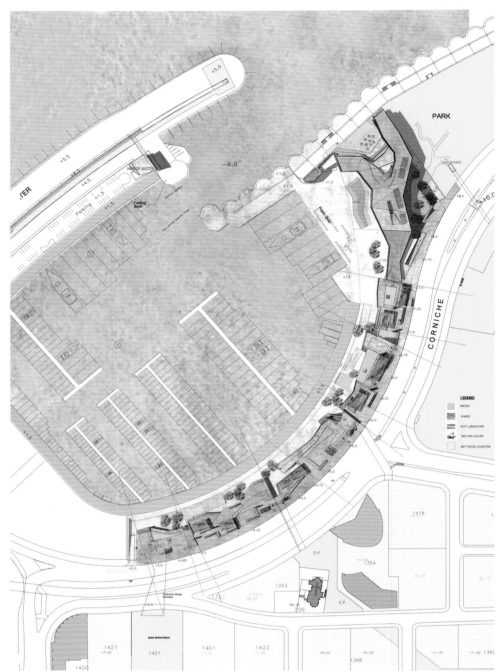

Site Plan
总平面图

Stairs and ramps are integrated to provide access to the waterfront level. The syncopated rhythm of platforms is achieved by constructing the overall curve of the Corniche in 5 angles related to the 5 reflection pools. Due to the variations in height along the Corniche, the platform levels and pools vary slightly in height allowing quiet, gravity-fed fountains to connect each pool level.

通过阶梯和坡道，游客可以直接来到水边游玩。这个平台的韵律感是通过整体结构 5 个不同角度的弧线形成的，这 5 个角度与 5 个泳池相关。因为高度的细微差别，这些平台层和泳池也稍许倾斜，从而构成一些安静且由重力产生的小瀑布，它们巧妙地连接不同的泳池。

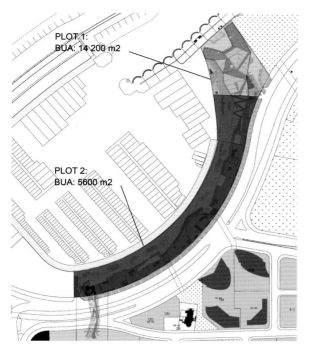

PLOT 1:
BUA: 14 200 m2

PLOT 2:
BUA: 5600 m2

Zones
区域图

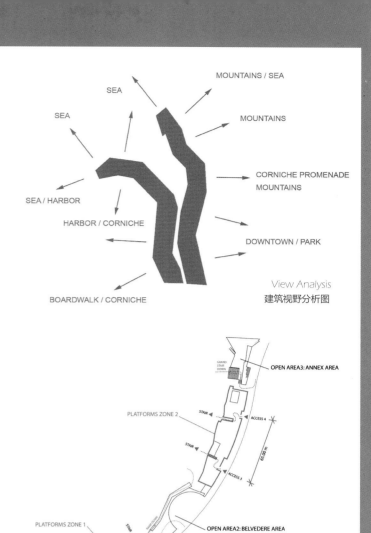

SEA

SEA

MOUNTAINS / SEA

MOUNTAINS

SEA / HARBOR

CORNICHE PROMENADE
MOUNTAINS

HARBOR / CORNICHE

DOWNTOWN / PARK

BOARDWALK / CORNICHE

View Analysis
建筑视野分析图

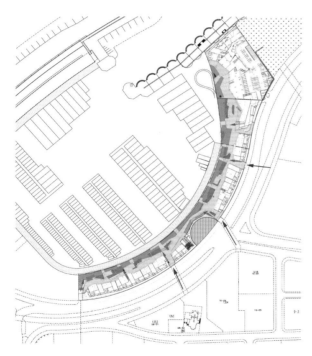

Lower Circulation
低层流线分析图

GRAND
STAIR
DOWN

OPEN AREA3: ANNEX AREA

PLATFORMS ZONE 2   STAIR   ACCESS 4

STAIR   ACCESS 3   65.00 m

PLATFORMS ZONE 1   STAIR

OPEN AREA2: BELVEDERE AREA

STAIR   ACCESS 2

STAIR DOWN   ACCESS 1   41.90m

OPEN AREA1: BRIDGE AREA

Platform Accesses Points
平台通道示意图

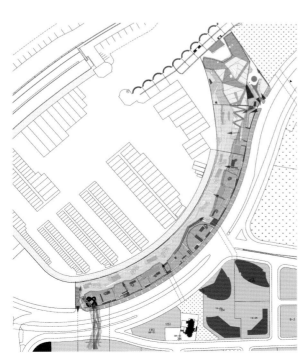

Upper Circulations
高层流线分析图

ARTIST AREA 5

5

ARTIST AREA 4

4

ARTIST AREA 3

3

ARTIST AREA 2

2

ARTIST AREA 1

1

Platform Diagram
平面示意图

SHAPE

The Beirut Marina Building takes its shape from strata and layers in forking vectors. Like the ancient beach that was once the site, the planar lapping waves of the sea inspire striated spaces in horizontal layers, as distinct from vertical objects. The horizontal and the planar become a geometric force shaping the new Harbor spaces. The form allows a striated organization of public and private spaces which include; restaurants and shops, public facilities, harbormaster, yacht club, and apartments above. The apartment building hinges to create a "y". The form produces a high ratio of exterior surface area offering a maximum amount of views, rising to form a public observation roof to the sea.

造型

项目的设计概念来自于分叉形式的层阶和叠层。如同古老的沙滩，海洋叠层的海浪激发设计师创造出这种分层的水平性的空间，它们与垂直的物体形成鲜明的对比。水平和二维的形式成了定义海港的新标准，这种形式使设计师能够形成分层的公共和私人空间的灵活组合，它包括餐厅、商店、公共设施、港务、游艇俱乐部、公寓等空间。公寓楼连接起来就像是一个"y"字。这样的造型提供了大片的室外区域，渐渐上升成为一片公共屋顶观景台，坐览广阔美景。

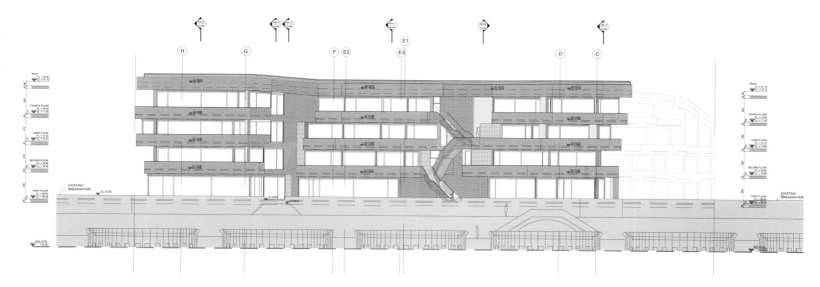

Elevation from Sea
海边方向立面图

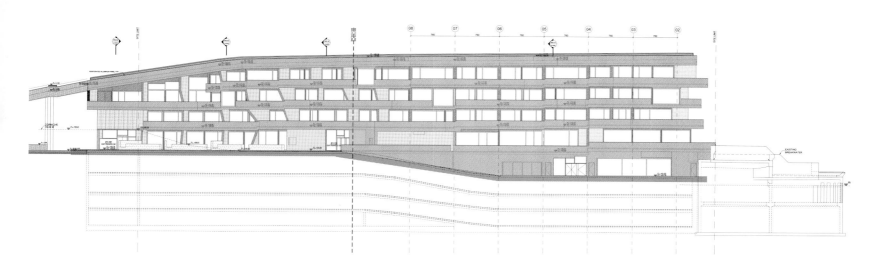

Elevation from Corniche
滨海路方向立面图

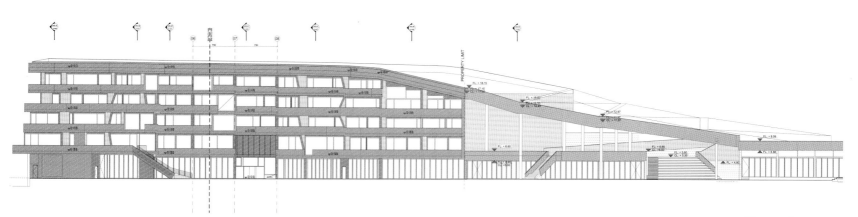

Elevation from Marina
码头方向立面图

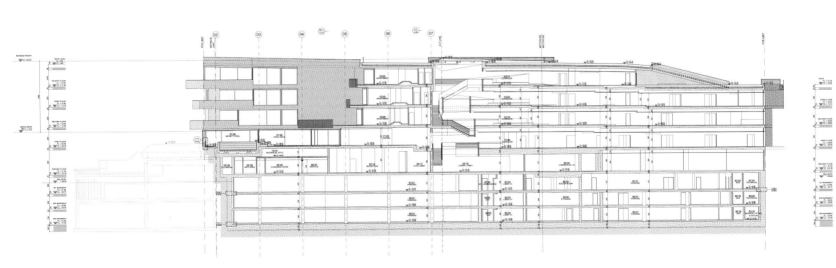

Longitudinal Section
纵向剖面图

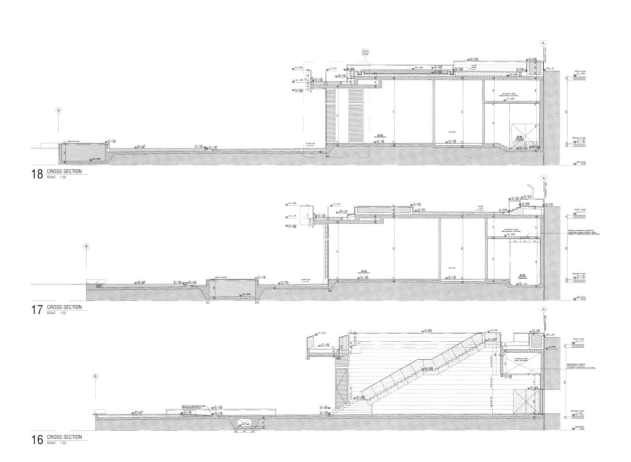

Cross Sections of Platform
露台横截面图

18 CROSS SECTION
SCALE 1:50

17 CROSS SECTION
SCALE 1:50

16 CROSS SECTION
SCALE 1:50

Ground Floor Plan
首层平面图

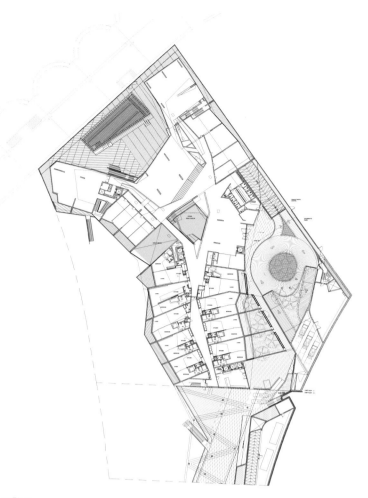

1st Floor Plan
1层平面图

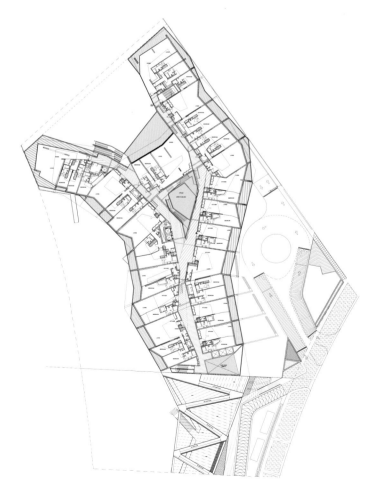

2nd Floor Plan
2层平面图

4th Floor Plan
4层平面图

Celebrating the sea horizon, the terraces are sculpted in local stone. The simple geometry of the upper platforms is in contrast to the colorful activities of restaurants below. The colored walls and fabrics below have their monochromatic and organic complements in awnings over the tables by the reflecting pools on the upper level. The building roof forms a public observation platform for the sea horizon.

海平线附近的台阶是用当地石材雕塑而成。上层平台简洁的几何造型和下层餐厅里丰富多彩的活动形成鲜明对比。彩色的墙壁和其下方的装饰物单色有机的组合和顶棚相呼应，并为上层水池边的桌子遮荫。建筑的屋顶是公共观景台，是欣赏海景的理想之处。

BALCONY 阳台
SUSTAINABILITY 可持续性
MATERIAL 材料
FAÇADE 立面

Block 20 Dwellings and Commercial, Barcelona, Spain
西班牙巴塞罗那 20 号街区商住两用公寓

Architect: NARCH (Joan Ramon Pascuets, Monica Mosset)
Client: rústic i natura s.l.
Location: Barcelona, Spain
Gross Floor Area: 3,500 m²
Floors: 6
Photography: Hisao Suzuki

设计公司：NARCH (Joan Ramon Pascuets, Monica Mosset)
客户：rústic i natura s.l.
地点：西班牙巴塞罗那市
建筑面积：3 500 平方米
层数：6
摄影：Hisao Suzuki

The building consists of a ground floor and five floors, for a 3 floors for commercial premises and the basement with 31 parking spaces and 20 storage rooms.

本项目包括首层和地上 5 层，其中 3 层作为商业用途，并配有地下室。地下室有 31 个停车位和 20 间仓库。

A light construction during the day is rich with dynamic volumes and shadows. At night, the building becomes like sifted light box, getting a fragile and porous building. The depth and variability by use of its inhabitants, create reflections, shadows and silhouettes that flow between interior and exterior. The building investigates the transparency and diversity. A thin perforated skin and a translucent veil work as a sunscreen and a visual filter to preserve the privacy of their occupants. The dwellings are arranged around the central core of communications and services, leaving all bedrooms in façade an all services inside. The location of the living rooms in the corner can play with the façade featuring balconies on either side free and making it disappear on the four planes of enclosure. 400 panels and perforated aluminum sheet: 3 mm anodized matte silver forms the second skin of the building, creating an intangible appearance and a porous architecture of imperceptible becoming. The panels are slided along the entire façades. The panels can be moved freely, getting an open system which varies according the necessity of the users and not determined by the architect.

Idea of Ecology

1) Lower costs associated with changing space configurations and greater design flexibility. Building is compact in shape to reduce their surface area.
2) Techniques and elements to reduce or maximize solar heat gain: sliding thin perforated anodized aluminium panels work as a second skin for the proper low-energy consumption.
3) Industrialized built process: pre-industrialized system based on panels in all façades.
4) Energy efficiency / low energy demands: materials, components and systems help to reduce energy consumption: big insulation, no-thermal pathways, energy efficient windows.
5) Improved occupant health and productivity taking advantage of cross ventilation and the use of solar gain.
6) Materials with long life expectancies reduce maintenance/replacement costs over the life of the building.

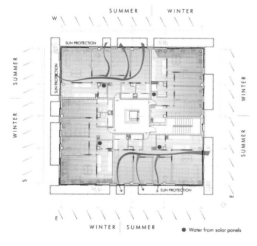

Green Materials

1) Recycled content: recycled wood sliding doors, porcelain tiles and EPDM waterproofing membranes come from recycling materials.
2) Locally available saving-energy resources for easy transportation: local brick, porcelain tiles...
3) Reusable or recyclable: aluminium panels, glass, cooper pipes, can easily dismantle and reuse at the end of their useful life.
4) Durable: 25 microns of anodised aluminium treatment façade.

Nature Resources

1) Renewable energies: flat solar collectors and vacuum for warming sanitary water. High-performance collectors allow annual savings between 50% to 70% of the required demand of energy required in a home, avoids almost a tonne of CO_2 per year.
2) Hot water storage. If the solar heat is not consumed immediately, you can store the solar and the heat can be used within hours of days for hot water or heating.
3) Condensing gas boiler: 98% efficient in energy exchanges, low CO_2 emission, classification nº5 in No_x emissions.
4) Each dwelling has natural cross ventilation. Mostly time is not necessary to use inverter air conditioning.

建筑在白天显得清淡雅致，动感的体量能产生丰富的光影效果。到了晚上，建筑又变成一个不断变化的灯箱，带来丰富多彩的变幻效果。根据住户使用方式的不同，各种深度和形式的变化让室内和室外之间形成映衬、阴影和轮廓的变化。建筑还注重透明度和多样性。多孔薄表皮和半透明覆盖层为建筑遮挡阳光，并作为视线过滤器，保护住户的隐私。所有的公寓都围绕着中央通信服务核心，而所有卧室都安置在立面的一侧，并且功能齐全。客厅则位于角落处，如此可和阳台互通，自由来往，400块面板和穿孔铝板：阳极氧化银是建筑的第二表皮。这种无形的表面和多孔结构让建筑显得更为细微。覆盖整个立面的面板可以滑动，打开可以扩展视野。设计师不做固定设计，住户可以根据需求而设置变化。

生态理念
1）空间配置的可变性和设计灵活度让建筑成本大大降低。紧凑的造型可以减少建筑的表面积。
2）采用技术和部件去减少或最大限度提高太阳能热增益：滑动阳极氧化铝多孔板作为第二表皮，能为建筑适当减少能量消耗。
3）工业化建筑过程：立面上的面板采用预制系统。
4 能源效率/低能源需求：材料、部件和系统能帮助减少能源消耗，如高度绝缘隔热走道、高效节能窗等。
5）利用交叉通风和太阳光来增强住户的健康和活动力。
6）选用寿命长的材料，减少维护成本，延长建筑的寿命。

绿色环保材料
1）循环材料：再生木推拉门、瓷砖、EPDM防水卷材都是利用可再生材料制成。
2）选用当地广泛使用且便于运输的节能材料和资源：本地砖、瓷砖等。
3）可重复使用或可回收材料：铝合金板、玻璃、铜水管，这些材料在使用寿命结束后可以轻易拆卸或再重复使用。
4）耐用：立面采用25微米厚的阳极氧化铝。

天然资源
1）可再生能源：平板太阳能集热器和真空饮用水加热器。高性能收集器每年可节省需求在一个家庭所需能量的50%~70%，并每年可减少一吨的了二氧化碳排放量。
2）热水存储。如果太阳能和热量还没有立即用完，住户可以将其存储起来，数小时或数天后再使用。
3）冷凝式燃气锅炉具有98%的能量交换效率，二氧化碳排放量低，并得到氮氧化物排放量nº5类别。
4）每间住宅都自然通风，大部分时间没有必要使用变频空调。

FEATURE 特点分析

SUSTAINABILITY

On the premier of meeting residential demands, the designer also pays attention to environmental protection and energy saving for sustainable development. Advanced technology, eco concept, green materials and recyclable resources are used in all aspects of the building, contributing it as a sustainable development model in the area.

可持续性

建筑在满足住宅功能需求的同时，更为注重环保节能的可持续发展要求。各种环保技术、生态理念、绿色材料和可回收资源被应用在建筑的各个方面，让建筑无愧成为本区可持续发展建筑的典范。

SHAPE 造型

BALCONY 阳台

MATERIAL 材料

SUSTAINABILITY 可持续性

Logements La Courrouze, Rennes, France

法国雷恩 Logements La Courrouze 公寓

Architect: Philippe Gazeau Architecte
Client: Territoires Publics
Location: Rennes, France
Site Area: 6,400 m²
Floors: 9
Photography: Philippe Ruault, Stephane Chalmeau

设计公司：Philippe Gazeau Architecte
客户：Territoires Publics
地点：法国雷恩市
占地面积：6 400 平方米
层数：9
摄影：Philippe Ruault、Stephane Chalmeau

The programme is located at the north-east end of the ZAC mixed development zone, in the "Bois Habité" area. It is bounded by the Rue Claude Bernard to the east, and by the Boulevard de Cleunay to the north.

本项目位于 ZAC 综合开发区的东部——Bois Habité 区。东边与 Rue Claude Bernard 路相接，北边则连通 de Cleunay 林荫大道。

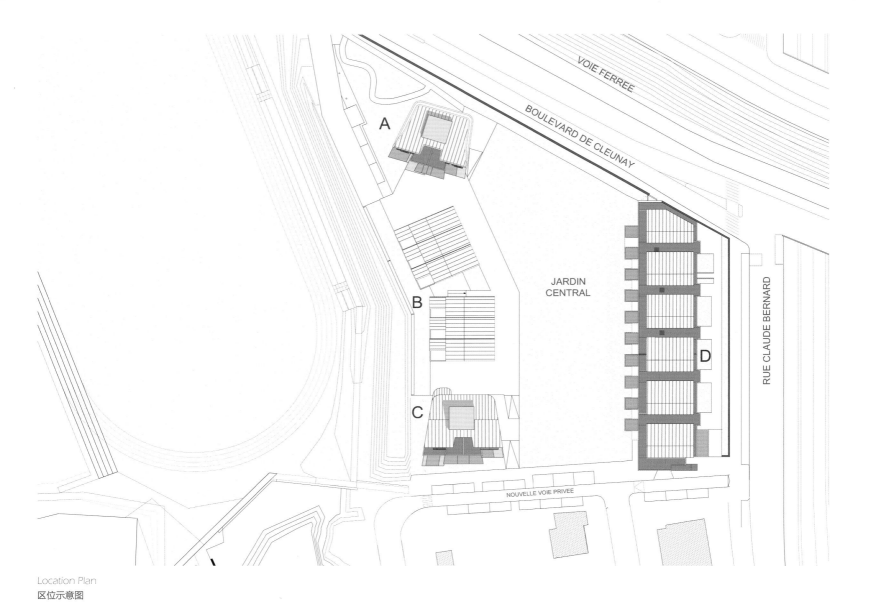

Location Plan
区位示意图

These architectural arrangements assert the desire to turn towards the city and its center, even though due to the orientation and the noise pollution from the railway line the north façade is less opened up and more height is given to the south façade, thereby giving the tower a rather unusual skyline. The tower's main body is wrapped on the north, east and west façades in a smooth, shiny mantle of vertical metallised ribbed cladding covering insulation on the outside of the building structure, in contrast with the more sculptural, mineral appearance of the south façade. On the ground floor, the tower fuselage is set on a brick base on the plaza ground or on the slightly sloping ground of the grassy areas.

The south tower block, which is lower and less slender in its proportions, has a dissymmetrical fuselage, thinner to the east on the inner garden side, broader to the west on the ZAC park side. This differentiation between the east and west façades is also emphasised in the color of the casing of vertical ribbed cladding which wraps around and insulates the north, east and west façades. The treatment of the south façade, and the ground floor base, are the same as on the first tower block. The rake on the fuselage is continuous all the way up to the sloping ridge. Any different treatment on the last two storeys would have had the effect of making the tower look stockier, being less tall than the other.

北塔楼是最高的一栋楼，特别是从雷恩市中心看过来时，让人感觉这是城市景观中最有气势的一栋建筑，这也是为什么顶层的造型——也就是最顶部两层楼的体量会特别不一般：相对主体量往里缩进，创造出一条护墙走道。这样的建筑布局是希望能面对城市中心，但由于这样会带来朝向问题和铁路的噪声污染，因此北立面很少打开，而南立面则相对较高，为建筑带来非同寻常的天际线风景。建筑主体的北、东、西立面被包裹在光滑、带有光泽的垂直金属棱纹覆层里，这层覆层将建筑的隔热结构也包裹在内，和南立面雕塑般的矿石材料外观形成对比。首层的体量设置在广场的砖石地面和略微倾斜的草地上。

南塔楼相对较矮，虽然比较细长但采用不对称造型，面对内花园的东侧较为细长，延伸到 ZAC 公园的西部。设计师采用具有隔热功能的垂直棱纹覆层覆盖北、东、西立面，以故意强调东、西立面之间的差异。南立面、首层基地的处理手法和第一栋楼相同。建筑倾斜的部分一直延续到坡地的高处。在最高两层楼做任何不同的处理都会使建筑看起来更厚实，和其他建筑相比也较矮。

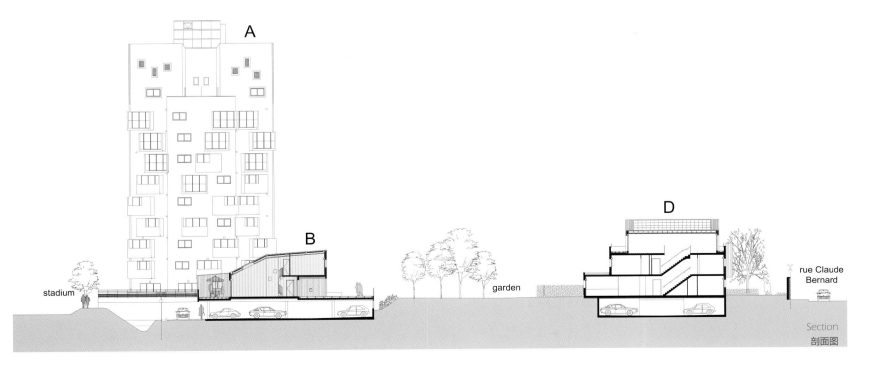

A

B

D

stadium

garden

rue Claude Bernard

Section

剖面图

BALCONY

As a contrast to the cool façade, the projected balconies with bright colors of green, yellow, pink, red inside soften the cold appearance of the building effectively. They also bring warm atmosphere into interior and also extend it to outdoor space. The bright colors will catch attentions even viewed from far distance.

阳台

与建筑体量冷色调表皮形成对比，每户阳台内侧的明亮色彩：绿、黄、粉、红等缓和了建筑的冰冷外观，将温暖的氛围带进室内，并延伸到室外空间。即使从远处看，阳台亮丽的色彩也能第一时间吸引眼球。

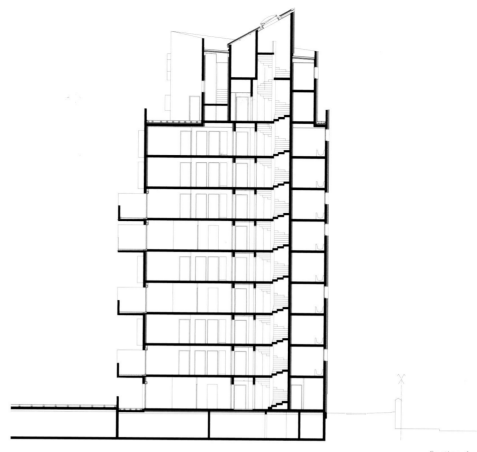

Section A
剖面图 A

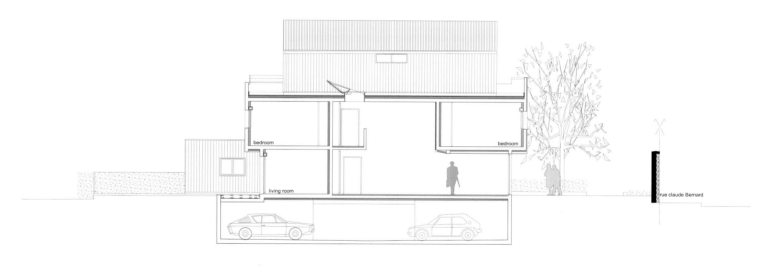

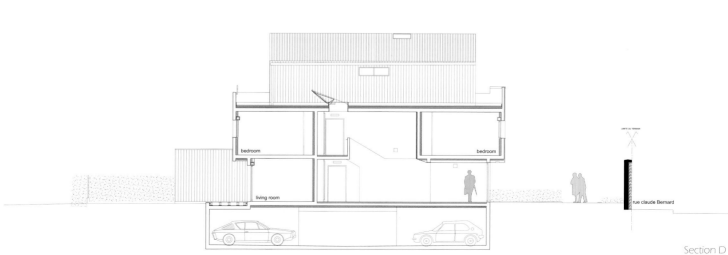

Section D
剖面图 D

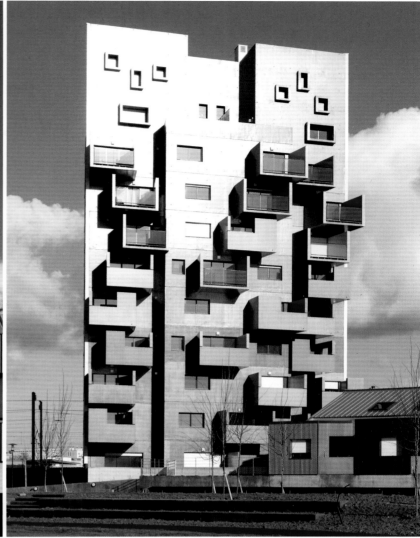

We chose to restrict the number of materials used, emphasized their aesthetics and durability, and their ability to blend together inside the operation and with their nearby urban setting, as well as their potential technically to meet the targets of passive insulation and long-term economy set for this project.

The colors associated with these materials tend to set them off against each other, and to emphasize to a greater or lesser degree certain special features linked to use (loggias) or to volumetric and architectural expression (the tower façades).

The façades on the north tower are covered on the north, east and west sides from the first to the eighth floors with vertical corrugated metallized cladding, and the same cladding for the last two floors. The ground floor base is lined with black brick.

The entire height of the south front is in surface coated concrete, its taupe color verging on rust, likewise on the outer faces of the terraces — projecting loggias. The inner faces of the loggias are in brighter colors, green, yellow, pink and red, to make these "outdoor rooms" feel more like part of the home. On the top floors the solid breast walls on the projecting boxes are replaced by colored glass guard rails.

For the cladding covering the south tower, the colors are treated bisymmetrically: dark green on the inner garden side, metal grey on the west side. The south facade is handled in the same spirit as the other tower.

The six housing units at the foot of the two towers are sheltered by surface coated concrete shells in the same color as the towers. Their sloping roofs consist of dark green steel deck.

The kitchens projecting out on the west side are covered with dark green and anthracite ribbed metal cladding.

设计师严格选择材料使用的量，注重满足美观性、耐用性和整体融合性，以及和周围城市环境的协调性，无源隔热的技术潜质满足项目的长期经济目标。

材料的颜色选择与它们之间的相互结合有一定联系，在不同的特殊功能使用上（阳台）、体量和建筑表达方式上（立面）强调或明或暗的效果。

北楼的北、东、西立面从 1 楼到 8 楼都覆盖了一层垂直波纹金属覆层，最后两层也采用了相同立面。首层地基则用黑砖作为内衬。

南立面的整个表面覆盖了混凝土层，采用带有锈色的灰褐色，凸出的阳台的表层也是如此。阳台的内部涂有明亮的色彩：绿、黄、粉、红，让这些户外空间更像是室内的一部分。顶层的楼层因为有结实的护壁，所以替代了凸出的阳台，换成彩色玻璃护栏。

南楼的覆层则采用了对称的色彩布局：面向内花园的一侧是深绿色，西侧则是金属灰。南立面也采用和其他塔楼相同的处理手法。

两栋楼首层的 6 个单元也是和塔楼同样颜色的混凝土覆层。房间的坡屋顶是用深绿色钢桥面构架而成。

在西面，向外凸出的厨房被涂成深绿色，并覆有无烟煤框架金属覆层。

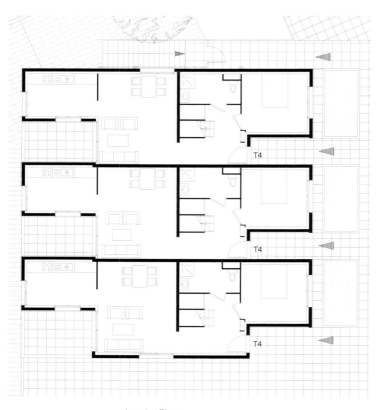

Interior Plan
室内平面图

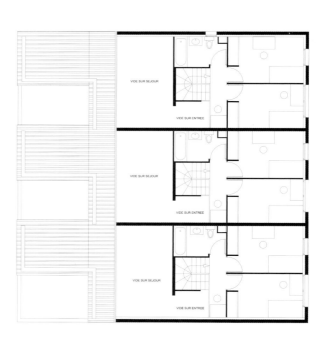

Floor Plan
楼层平面图

T4   T2   T3

13P

0 1M 2 3 4 5

PLAN ETAGE COURANT

COMBLE   COMBLE

PLAN R+7

T5   T5

PLAN R+6

T4   T2

0 1M 2 3 4 5   PLAN REZ DE CHAUSSEE

Unit Plan
户型图

SUSTAINABILITY 可持续性

FAÇADE 立面

OPEN SPACE 开放空间

MATERIAL 材料

"The Writers", London, UK

英国伦敦 "作家" 综合大厦

Architect: ORMS Architecture Design
Client: Artillery Lane Ltd
Location: London, UK
Gross Floor Area: 2,322.6 m²
Floors: 5, 6
Photography: Edward Hill

设计公司：ORMS Architecture Design
客户：Artillery Lane Ltd.
地点：英国伦敦市
建筑面积：2 322.6 平方米
层数：5、6
摄影：Edward Hill

ORMS Architecture Design completed the transformation of two office buildings into a stylish residential complex, united by a tranquil central courtyard in the city of London. The project for Artillery Lane Ltd. includes the refurbishment of a total 2,322.6 m² office accommodation to provide 26 apartments and 743.2 m² retail and restaurant space at ground and basement levels.

Taking its name from the famous nearby market, "The Writers" project is located within easy reach of Liverpool Street Station and lies within the Bishopsgate Conservation Area. By sensitively retaining and extending the street-facing façades of the original office buildings on Middlesex Street and Artillery Lane, the architect has created a reinvigorated building which enhances the existing streetscape of the Conservation Area. Once visitors are within the courtyard, its open, contemporary face is revealed.

本次设计师在伦敦设计的项目是将原来的两栋办公大楼转变为时尚住宅综合楼，联系两栋楼的是一座宁静的中央庭院。项目的翻新内容包括将占地 2 322.6 平方米的办公楼改造成拥有 26 间公寓、占地 743.2 平方米的首层零售和餐饮空间以及地下室楼层。

项目的名字来自附近著名的市场，"作家" 综合大厦位于毕晓普保护区内，从这里可以轻松前往利物浦车站。设计师小心地保留并延伸了朝向 Middlesex 街和 Artillery Lane 公司大楼的原办公楼立面，将建筑重新改造，使之富有活力，更是成为了保护区的一景。来客一旦走进中庭，现代的气息随即迎面而来。

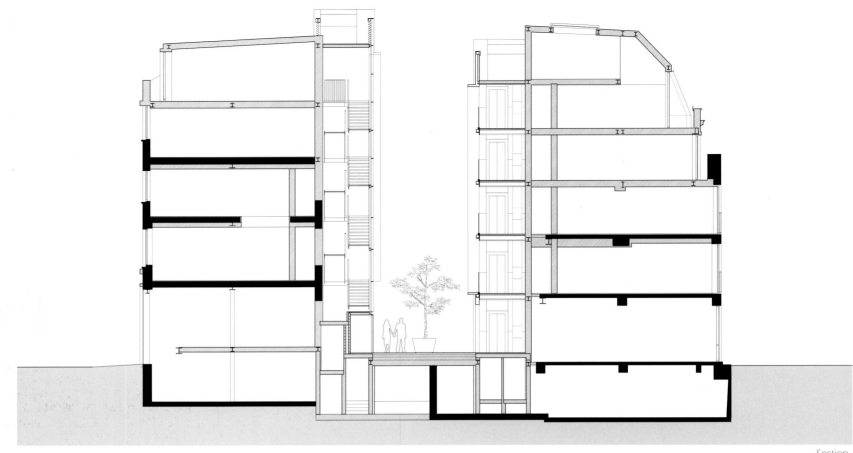

Section
剖面图

The yellow stock brickwork of the façade of the Middlesex Street building has been sensitively restored and a new brickwork parapet was added, to tie in with the cornice height of the adjacent buildings. New double glazed windows with a central spandrel panel increase the vertical reading of the façade, and a new zinc clad top floor has replaced the original pitched roof. The proportions of the fenestration were derived from an analysis of the adjacent window proportions.

From street level it is not immediately apparent that the Artillery Lane building has been extensively refurbished and extended. The façade has been restored along with the formation of new openings at either end of the elevation, and new central Juliette balconies were added to the center. Traditional timber sash and case windows with high performing double glazing have replaced the original windows but it is the new three storey extension on top that has been purposely designed to minimise its impact from the street level.

At ground level, the façade has been "opened up" to increase views into the restaurant with a new rendered colonnade added to frame the windows and "ground" the building. The refurbished building now sits sensitively within the surrounding context.

The new third floor is set back from the street and constructed to use a traditional brick cavity wall with a further two storeys added on top in a zinc clad rooftop extension. The rooftop is a mansard and sets even further back from the third floor, hence it is not visible from street level. The new extensions are formed from lightweight and sustainable FJI joists which are packed with insulation to form a highly insulated envelope.

建筑面向 Middlesex 街的黄色砖砌立面被精心修复，并加建了一面砖砌护墙，以配合相邻建筑的檐口高度。新的双层玻璃窗和中央拱肩板提升了立面的垂直感，另外新建的锌复合顶层取代了原来的斜屋顶。开窗的设置则是根据相邻窗口的比例分析而定。

从街道上看的话，行人并不能马上觉察出建筑已经经过全面的翻新和扩建。两端新开口形成的立面部分被修复，英伦风情阳台也被加入到中央庭院中。传统的木饰、窗框、高性能双层玻璃取代了原来的窗口，位于新扩建的三层之上，这样设计的目的是为了将建筑对街道的影响减到最小。

首层的立面被"打开"，为餐厅扩宽视野；新增加的柱廊也成为了窗外和首层的一景。经过翻新的建筑如今和周围环境相得益彰。

新建的第三层从临街面往后缩进，结构是传统砖砌空心墙。再往上两层楼，顶部就是扩建的锌复合屋顶。屋顶是双层斜坡结构，相比起第三层更为缩进，因此从街道上是看不到屋顶的。新的扩展部分使用轻质、可持续的 FJI 材料，附带的隔热功能让整个立面都具有高度隔热功效。

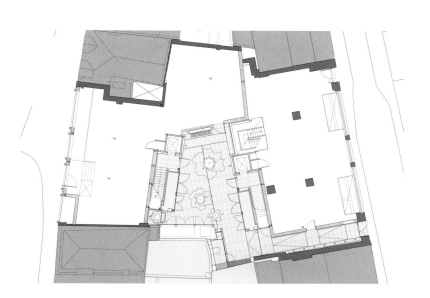

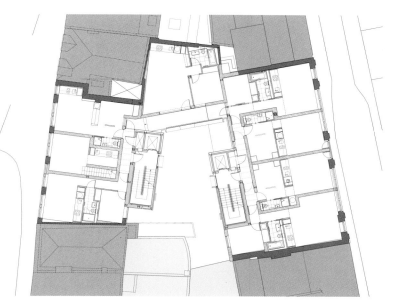

Ground Floor Plan
首层平面图

2nd Floor Plan
2 层平面图

MATERIAL 材料
FAÇADE 立面
SHADE 遮阳
SUSTAINABILITY 可持续性

Shelford Road Apartments, Singapore

新加坡赐福路公寓

Architect: Forum Architects
location: Singapore
Floors: 5
Photography: Albert Lim

设计公司：Forum Architects
地点： 新加坡
层数：5
摄影：Albert Lim

The site is situated in a prime residential area on the fringe of the city. The site dips suddenly to a flat area four storeys below the road. The nature of the site whilst presenting access and construction difficulties, creates a unique opportunity for an exclusive retreat nestled away from the city, barely five minutes away.

本案位于城市边缘的黄金居住区。该处地形陡然下降到街道平面以下的平坦区域，此处建筑的首4层楼都位于街道的下方。场地的自然地形虽然为入口通道和建设施工带来了难题，但也为项目提供了绝佳的位置，远离又临近着城市，从这里到城市的距离只有五分钟的路程。

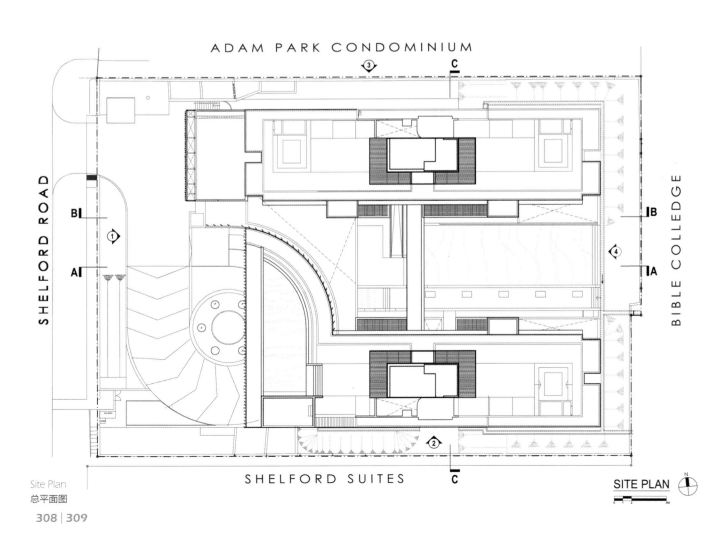

ADAM PARK CONDOMINIUM

SHELFORD ROAD

BIBLE COLLEDGE

SHELFORD SUITES

SITE PLAN

Site Plan
总平面图

An urban screened wall facing the road belies the secluded, private and cosy feel of the residential grounds. The two blocks are connected via a 15.7 m slender bridge, yet have another threshold to guard privacy of the residents.

Oriented North-South to minimize solar heat gain in the tropical climate, two rectangular blocks, housing fifteen residential units and four penthouses enclose a sunken private courtyard with a 21.5 m swimming pool.

一面面向街道的屏蔽墙为居民带来了幽静、隐私和舒适感。两栋建筑间有一座长 15.7 米的长桥连接，另有一个入口来保护居民的隐私。

建筑坐北朝南，以尽量减少热带气候的太阳辐射。项目由两栋长方形体量组成，拥有 15 间公寓房、4 间复式套房、下沉的私人庭院和一个长 21.5 米的泳池。

FEATURE 特点分析

FAÇADE

Full height glass façade, carefully oriented north and south away from the direct sun, adorn the living space, brings natural light and affords unimpeded view. The glass doors slide open to create a 3.6 m-wide clear opening to allow the units to be naturally cross ventilated and transforming the living and dining room into semi-outdoor spaces, especially for units at the level of the swimming pool.

立面

建筑被全高的玻璃幕墙覆盖，坐北朝南以避免阳光直射，并为生活空间带来自然光和开阔视野。玻璃门宽 3.6 米，巨大的开口方便室内的自然交叉通风，并将客厅和餐厅变成半户外空间，特别对游泳池同一层的住户效果更佳。

Residents and visitors "walk on water" through the spacious lobby to enter their units. The movement of water is projected onto the ceiling as natural light is reflected off the reflective pool's surface, creating a subtle montage of overlapping lights and shadows.

Residents who drive access to the basement carpark five storeys below the road via a ramp, lit by skylights under a reflective pool, surrounded by a bamboo screen on the outside of the ramp.

The penthouse units have rooftop Jacuzzis and pools to take advantage of the green setting.

居民和来客穿过水上通路，来到宽敞的大厅，进而进入各自的房间。水流的波纹被反射到天花板上，自然光被水面反射，从而产生了微妙的蒙太奇式光线和阴影的重叠。

居民驾车穿过坡道，即可来到低于街道面的5层地下停车场。位于水池下方的天窗为停车场引进光线，周围有竹林围绕，以屏蔽外部坡道外的噪声。

复式套房利用绿色环境的优势，设置了按摩浴缸和泳池。

Specially designed triangular screens are provided at the master baths as well as the west facing façades. The triangular profile of each screen member creates multiple colors as a result of light and shadow, whilst allowing the degree of privacy to be varied according to how the members are rotated.

特别设计的三角屏风被放置在主浴室里和西向的立面上，它们不但创造了多彩的光影效果，而且根据旋转的角度不同产生不同程度的隐私保护效果。

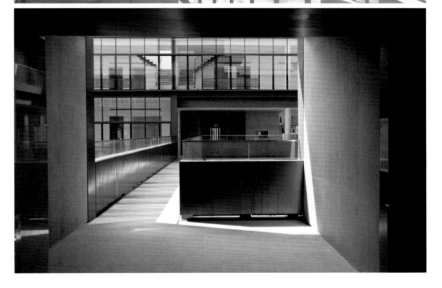

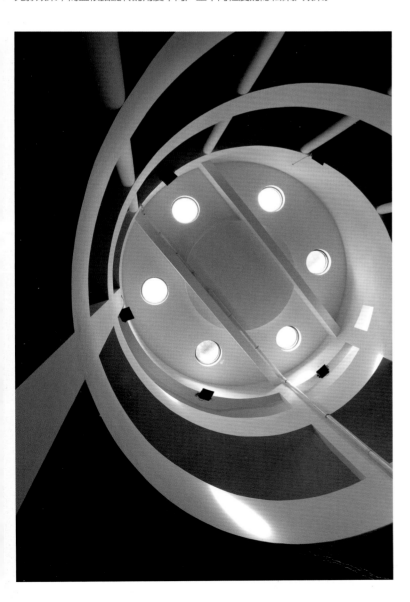

SHADE 遮阳
FAÇADE 立面
COLOR 色彩
MATERIAL 材料

Block A, Amsterdam, the Netherlands

荷兰阿姆斯特丹 A 住宅大楼

Architect: Dick van Gameren Architecten
Client: Far West / De Principaal, Amsterdam
Location: Amsterdam, the Netherlands
Floors: 13
Photography: Marcel van der Burg - Primabeeld

设计公司：Dick van Gameren Architecten
客户：Far West / De Principaal, Amsterdam
地点：荷兰阿姆斯特丹市
层数：13
摄影：Marcel van der Burg — Primabeeld

Block A is the first of three new blocks that are being built in the north of Delflandplein street in the Amsterdam district of Slotervaart.

Maximum density was strived for in order to set a large-scale renovation of the Delflandplein neighborhood into motion for the first new block of buildings: 170 social council dwellings, a day activity center and business spaces will replace the original building, three strips of housing only several storeys high.

阿姆斯特丹 Slotervaart 区北部 Delflandplein 街区正在建设 3 栋新大楼，本项目就是其中最先建设的一栋。

Delflandplein 街区将进行大规模的整修，本项目作为第一栋新大楼，设计师希望能争取到最大化密度。建筑包括 170 间社会住宅、一个日间活动中心和商业空间，这些设置将取代原有 3 栋只有 7 层高的建筑。

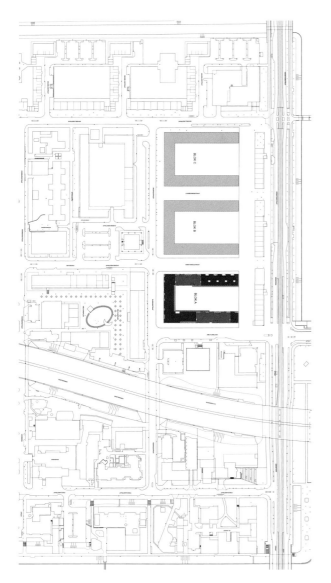

Location Plan
区位示意图

The largest possible variety of housing types and sizes were built in the block, ranging from small studios with common living rooms for people in need of care to relatively spacious six-room maisonettes and large five-room apartments. The trend towards reducing the programme for stacked dwellings to average three-room apartments of 90 m² could thus be avoided. As a result, it was possible to build multiple types of residences in this block, and the original ideals of the western suburbs — a balanced differentiation of dwellings for the elderly, starters and families — have been restored here in an updated form.

建筑的住房有各种户型和大小，从小型工作室或为需要照顾家人而设计的公用客厅，到宽敞的六室豪宅或五室公寓都一应俱全。普通的住宅为了减少用地，会将住宅空间层叠，形成 90 平方米的三室单元，而本案的设计师却希望避免这种布局。于是，建筑设置了多种类型的住宅，而最初在西郊区建设更多不同类型的住宅也能得到实现，为老人、新婚夫妇和大家庭提供更多选择。

A   WiBo (Living in Protected Environment) Houses
B   Residential Care with common living rooms
C   Day Activity Center
D   Apartments
E   Maisonnette Family Houses
F   Elderly Residential Group

Functions Analysis
功能分析图

Section
剖面图

An inner garden is being created on a car park in the heart of the building. Earthen hills make it possible to place large trees in the grass.

The considerable noise pollution from the nearby A10 Amsterdam Ring Road was the reason for this particular façade design. The two higher parts of the U-shaped block have a very closed façade facing the Ring Road. The opposite west side, on the other hand, was made as open as possible with large frontages and outer areas.

在建筑的中心停车场上还有一座内部花园。土丘上可以种植草坪和大型树木。

A10 阿姆斯特丹环城路是附近最大的噪声污染源，建筑因此采用特殊的立面设计。U 形体量的两个高出部分面向环城路，因此立面是封闭的。而西侧的立面则是尽量打开，向街道和外部空间开放。

FEATURE 特点分析

FAÇADE

The building inherits traditional architectural technology of Slotervaar Region while tries to break the monotonous, repetitive patterns by a woven-carpet-like façade, bringing an impressive to the neighbourhood. The woven carpets each spaning on two floors cover the whole façade, forming an integrated crisscross pattern that endows the building with a sculptural and robust appearance.

立面

建筑的立面传承了 Slotervaar 地区的古老建筑技术，却又打破其单调、重复的模式，用编织毯图案组成的立面给街区带来焕然一新的感觉。编织毯元素覆盖整个立面，每块都横跨两层楼，且相互之间的图案纵横交错，赋予建筑一种雕塑感和强有力的外观。

The large prefab concrete façade panels, containing brickwork, allude to the original brick assembly technique used in Slotervaart, but at the same time attempt to break with the monotonous, repetitive appearance of traditional prefab façades.

Due to the different tints of stone, as well as the alternating relation of the surface of the stones and the joints in the different brickwork patterns, a relatively colorful image emerges like woven carpets hanging next to and above each other.

大型预制混凝土立面板，其中含有砖墙结构，这暗喻了 Slotervaar 地区古老的砖组装技术，但同时设计师也尝试打破传统的单调、重复的立面。

因为石材的不同色彩，以及立面石材和接口处的砖砌图案之间的相互联系，如同编织毯般相互交错的多彩立面呈现给大家。

Ground Floor Plan
首层平面图

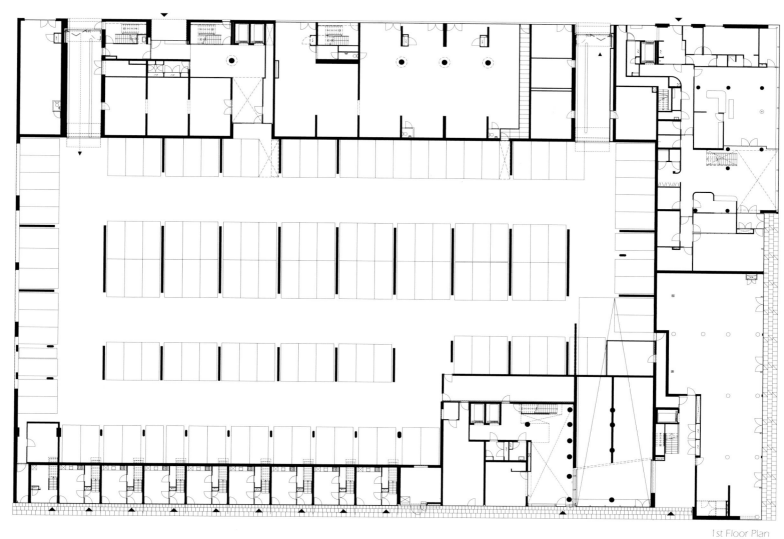

1st Floor Plan
1 层平面图

5th Floor Plan
5 层平面图

FAÇADE 立面

COLOR 色彩

SHAPE 造型

SUSTAINABILITY 可持续性

Brandon Street Affordable Housing, London, UK

英国伦敦布兰顿街经济住房

Architect: Metaphorm Architects
Location: London, UK
Client: Southwark Council, London & Quadrant
Floors: 5

设计公司：Metaphorm Architects
地点：英国伦敦市
客户：南华克市政府、伦敦和区域集团
层数：5

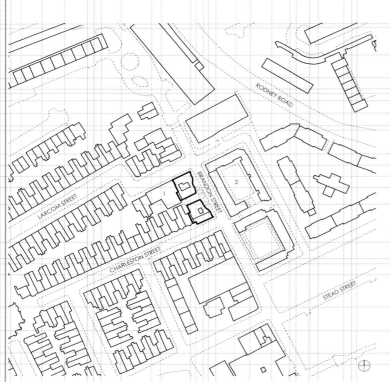

Won through an architectural competition, the Elephant & Castle Regeneration housing project for Brandon Street in central London is part of Southwark Council District government's plan to establish architectural quality benchmarks for the Borough. The Council's aim was to remove differences between public and private housing and to start to evolve a new generation of residential architecture for Southwark which expresses itself in terms of an urban form which is contemporary, light, flexible, mixed tenure and contained within a mixed-use environment.

本案位于伦敦市中心布兰顿街，于建筑设计比赛中胜出，是 Elephant & Castle 城市再生住房计划的一部分。该计划由南华克区政府发起，旨在提高市区的建筑质量。政府的目的是希望消除公用和私人住宅间的差距，建造南华克区新一代住宅建筑，在综合城市环境中打造现代、轻质、灵活、可持续的城市建筑。

Location Plan
区位图

FEATURE 特点分析

COLOR

Different from formal façade of general architecture, the façade of the project uses a kind of gradient color from the honey yellow to Bordeaux red, brings a warm and pleasant sense for people. Covered by hexagonal tile cladding, the building is like a fish swimming leisurely in the green city with its curve shape.

色彩

与一般建筑的普遍立面不同，本案建筑的立面采用暖色渐变手法，从蜜黄色到枣红色的渐变能给人带来温暖和愉悦的感觉。立面覆层是六边形的瓷砖覆层，配合建筑的曲线造型，仿佛是一条大鱼游曳在绿色城区中，颇为有趣。

Metaphorm Architects' response was, despite high density requirements, to create urbanity through public open space, to initiate and acknowledge Brandon Street's increased future significance as an axis into the regenerated Elephant & Castle, to create characterful, yet practical dwellings, and to retain a number of existing trees.

The black brick side elevations are a contemporary extension of the established material pattern along adjoining roads. Ceramic-clad in 37 gradient color tones, ranging from honey yellow to bordeaux red, the main façade breaks this continuity through a strong contrast, announcing the changing nature of Brandon Street and creating an element of delectation. Meandering past the existing trees, it defines the character of both, external public and internal private spaces. The grey-white dual toned pre-cast concrete benches following the curvilinear façade are an invitation to stay, extended in particular to the children of the pre-school adjoining and to the elderly of the almshouse opposite.

Urbanistic considerations led to the creation of two blocks, each five-storey high, separated through a small private courtyard. The orientation of the apartments varies between ground floor and upper floors. By not requiring windows on the eastern façade, this layout also permits the creation a public space which does not interfere with the privacy of the ground floor units.

The use of an off-site manufactured Light-Gauge-Steel structure for tight-radius undulating walls is unprecedented, and the architectural design had to provide precisely determined variable pre-fabricated wall panel widths, as a function of radii, window positions, minimum and maximum wall cavity widths, etc, to avoid faceting of the façade. The building envelope comprises walls of inner leaf LGS panels with integral insulation and outer leaves of masonry or render on insulation, resulting in a compact wall construction achieving U-Values as low as $0.11 W/m^2$, a PassivHaus level, within an average-sized wall depth. The scheme achieves BREEAM Code for Sustainable Homes Level 4.

设计师对此规划的回应是尽管住宅密度需求高，仍然致力建造温文尔雅的城市开放空间，为布兰顿街的改建和将来发展增添一座地标。建筑开始确立布兰顿街为贯穿过 Elephant & Castle 区的轴线，为居民提供特色而实用的住宅单元，同时还保留了原有的树木。

青砖侧立面是毗邻街道上的材料和图案的现代延伸。主立面瓷砖的颜色有 37 种渐变色调，从蜜黄色到枣红色，让主立面以强烈的对比打破一贯的连续性，宣告布兰顿街的本质变化，并创造欢愉的元素。建筑在保留下来的树丛中蜿蜒，定义出外部公共空间和内部私人空间。曲线立面前的灰白双色调预制混凝土长椅仿佛在邀请行人坐下，特别为幼儿园儿童和对面救济院的老人们服务。

考虑到都市化因素，项目分为两栋楼，各高 5 层，中间间隔一座私人小庭院。首层和高层单元的朝向都各不相同。因为东立面上不需要窗口，因此得出的布局可以缔造出一个公共空间，同时不会干扰到首层单元的隐私。

使用场外制造的轻量钢结构作为沿着小半径波浪起伏的墙壁是前所未有的手法，同时建筑设计必须提供精确的预制墙板宽度的变量、半径的函数、窗口位置、最小或最大的壁腔宽度等，以避免影响到立面。建筑的立面包括具有隔热功能的 LGS 板内墙和砖石，或饰面砂浆隔热板外层，如此可形成紧凑的墙体结构，并在一般墙体深度内将 U 值降低至德国 PassivHaus 标准的 0.11W/ m²。本案获得了 BREEAM 可持续住宅规范等级 4 级认证。

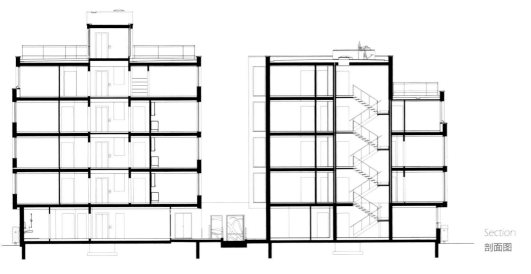

Section
剖面图

BRANDON STREET

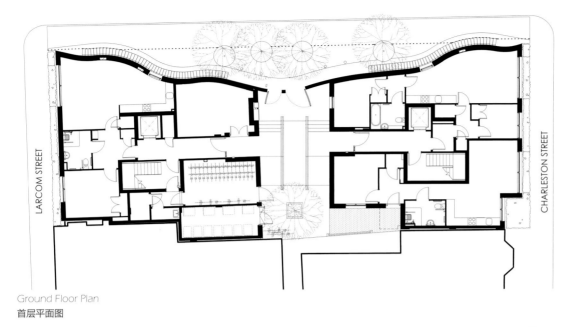

LARCOM STREET

CHARLESTON STREET

Ground Floor Plan
首层平面图

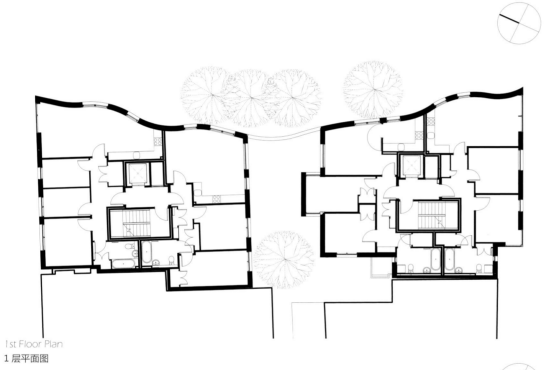

1st Floor Plan
1 层平面图

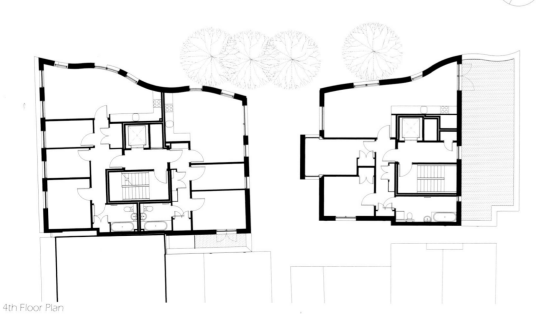

4th Floor Plan
4 层平面图

WINDOW 窗
FAÇADE 立面
LAYOUT 布局
STRUCTURE 结构

The Fuglsang Lake Centre, Herning, Denmark
丹麦海宁福格桑湖畔中心

Architect: C. F. Møller Architects
Client: Herning Municipality
Location: Fuglsang Toft, Herning, Denmark
Gross Floor Area: 11,000 m²
Floors: 5
Realization: 2012

设计公司：C. F. Møller Architects
客户：海宁市政府
地点：丹麦海宁福格桑区
建筑面积：11 000 平方米
层数：5
完成时间：2012 年

The centre is located on the banks of Fuglsang Lake. This unique location — together with all the common facilities, the centre will provide for the area's elderly, including activities rooms, an IT room, workshops, a hairdresser, shop, gym, clinic and library — has inspired the name of the project, the Town by the Lake.

本项目坐落在福格桑湖边，占据优越的地理位置。中心将会和其他公共设施一起，为该片区的老人提供各种活动场所，包括活动室、IT 室、工作室、理发店、商店、健身房、诊所和图书馆等。项目的命名也由此得来——湖边的市中心。

FAÇADE

Via a transparent façade, the landscape is drawn into the urban environment of the ground floor, which, with its many niches, plazas, provides rooms for all kinds of social interaction from the community as a whole to intimate groups or solitary reflection.

立面

透过透明的立面，景观被引入建筑首层的城市环境中，首层的商店、广场为各种社交活动提供场所，社区、群体或个人活动都能得到满足。

Site Plan
总平面图

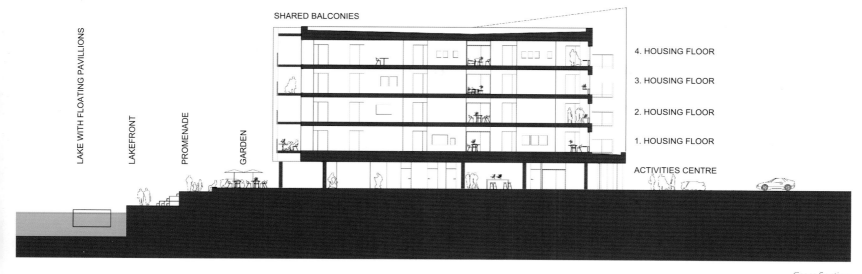

SHARED BALCONIES

LAKE WITH FLOATING PAVILLIONS
LAKEFRONT
PROMENADE
GARDEN

4. HOUSING FLOOR

3. HOUSING FLOOR

2. HOUSING FLOOR

1. HOUSING FLOOR

ACTIVITIES CENTRE

Cross Section
横截面图

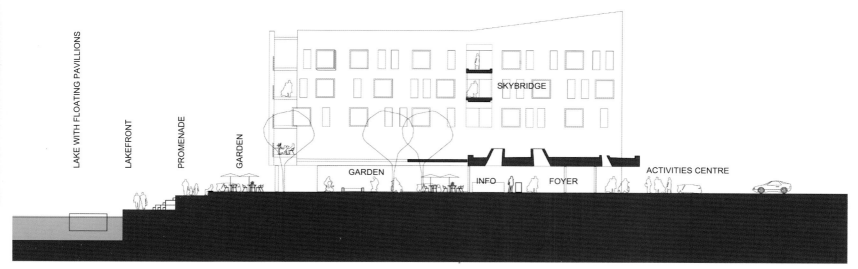

LAKE WITH FLOATING PAVILLIONS
LAKEFRONT
PROMENADE
GARDEN

SKYBRIDGE

GARDEN

INFO   FOYER

ACTIVITIES CENTRE

Cross Section
横截面图

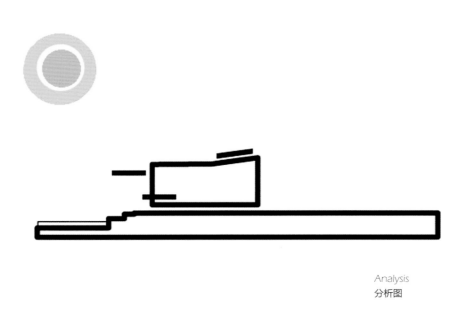

Analysis
分析图

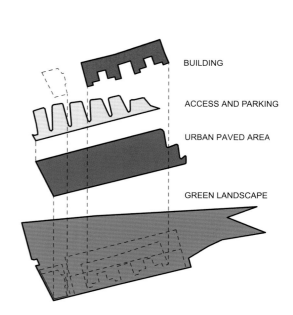

BUILDING

ACCESS AND PARKING

URBAN PAVED AREA

GREEN LANDSCAPE

Diagram Layers
布局示意图

East Elevation
东立面图

North Elevation
北立面图

Sorth Elevation
南立面图

The centre, with a total of 112 units, is oriented towards the lake, in the shape of a comb with its teeth towards the water. The centre is structured like a town, with urban elements such as a main street, neighborhoods and gardens. Along the main street are the town's four neighborhoods, housing the various functions: day centre, physical training area, activities area, a café and shops. The gardens lie between the buildings, which offers sensory experiences, including a vegetable garden and chicken-keeping park, and conclude with sitting steps at the quayside by the lake. Above the "town" — in up to three and four storeys — the individual residences comprise small and intimate units in groups of twice eight. The residences are linked to the urban area via small plaza areas on the main street.

The possibility of being able to put your personal touch on your home has been given a high priority. For example, all of the apartments have their own bay windows which residents can arrange in their own way. At the same time, the bays also bring dynamics and character to the centre's light rendered façades.

中心面朝湖边，共有 112 间住房，外形像是一把梳子，梳齿部分朝向湖边。项目的结构就像是一座小城镇，一般的大道、街区和花园都一应俱全。沿着主干道分布了 4 个街区，分别容纳了不同的功能区：日间中心、体能训练区、活动区、咖啡厅和商店等。花园位于建筑与建筑之间，为人们带来感官上的享受。菜园、饲养园，湖边码头还有台阶供人闲坐。"城镇"的上层空间则是 3~4 高层的住宅楼，每栋都有小型的私人房间，每 16 间组成一个组团。住宅楼通过市区的主要街道上和各个小广场联系在一起。

公寓的设置注重灵活性，居民可以按照自家的喜好布置，如所有的公寓都有自己的凸窗，居民可以自行决定使用方式。同时，海湾为项目立面的光照效果带来了动感和特色。

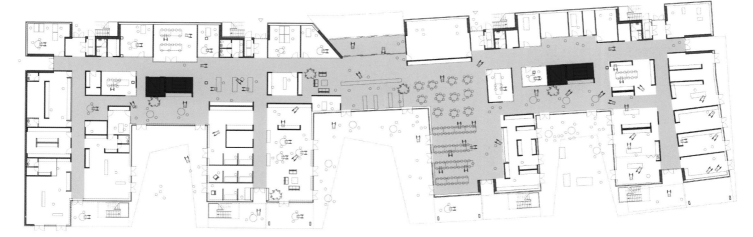

Ground Floor Plan
首层平面图

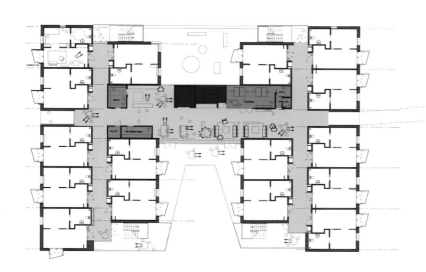

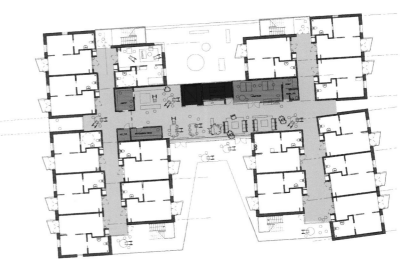

Upper Floor Plan
上层楼层平面图

Living Units Detail Plan
居住单元细节规划图

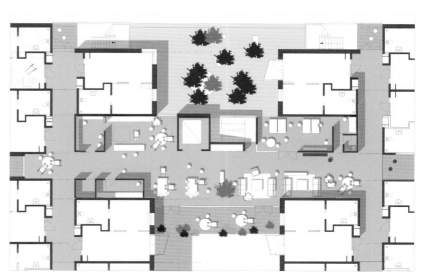

Common Areas Detail Plan
公共空间细节规划图

FAÇADE 立面
STRUCTURE 结构
MATERIAL 材料
VIEW 视野

Habitations Rue Pierre-Rebière, Paris, France

法国巴黎 Rue Pierre-Rebière 街公寓

Architect: Agence Christophe Rousselle, Agence Nicolas Laisné
Location: Paris, France
Site Area: 980 m² (Black building), 577 m² (Wooden building)
Floors: 7 (Black building), 4 (Wooden building)
Photography: Philippe Ruault, Nicolas Laisné

设计公司：Agence Christophe Rousselle、Agence Nicolas Laisné
地点：法国巴黎市
占地面积：980 平方米（黑色建筑），577 平方米（木质建筑）
层数：7（黑色建筑），4（木质建筑）
摄影：Philippe Ruault、Nicolas Laisné

These two buildings housing 22 apartments is situated in the Porte de Clichy situated in the north-west of Paris. Rue Pierre-Rebière is a straight road measuring 600 m in length and 25 m in width. It is bounded by the Batignolles Cemetery and the grounds of the Honoré-de-Balzac High School. This narrow and once neglected street will be transformed by a host of new structures.

本案由两栋住宅大楼组成，共有 22 间公寓，位于巴黎 西 北 的 Porte de Clichy 区。Rue Pierre-Rebière 街是一条直路，长 600 米，宽 25 米。两头分别连接 Batignolles 公墓和 Honoré-de-Balzac 高中操场。一些新的建筑将为这条狭窄又冷清的街道增添新的活力。

STRUCTURE

The structure of the black building is formed by staggered and stacked floors from different directions, irregularity but orderly. These protruding floors in each direction well avoid lighting shielded at lower floors by higher floors and offer a close contact to the sky. The wooden building is partly outward extended in a boldly design for its limited site to larger available spaces for residents. There are also lots of windows opened on the façade to compensate natural lighting for its closure structure.

结构

黑色建筑的结构由从不同方向错开突出楼层的组成，参差不齐中却又井然有序。往各个方向凸出的楼层避免了上层遮挡下层采光的问题，并提供给住户与天空近距离接触的机会；木质建筑由于场地狭小，设计师大胆地建造向外延伸结构，以加大住户可使用的面积，并在各立面设置许多窗口，以弥补建筑封闭结构的采光问题。

Different yet complementary, the two buildings create a dialogue and give a staccato-like rhythm to the street. The first building projects horizontal lines, while the second is resolutely vertical and seems to reach to the treetops. Should passersby care to look, they will be able to glimpse the trees of the cemetery through the space left between the buildings. The apartments in both buildings have generous balconies and the planting creates a living link between the street and the greenery of the cemetery. The apartments face the street and the gardens and are protected from any noise originating on the cemetery side. The homes on the ground floor all have access to well-oriented private gardens (south–south west – south east), which act as a filter in terms of the street. Each building has its own entrance and residents cross a small planted area to reach the entrance halls.

两栋建筑相互不同又相互弥补，形成了一种完美的互动，同时呈现出一种不连续的节奏感。第一座建筑打造水平的线条，而第二座建筑则是完全纵向的，看起来好像要超越树端。如果行人从周边经过，他们能从两座建筑之间的缝隙看到附近墓地的绿树。两座建筑的公寓都有开阔的阳台，而种植的绿色植被则连接了街道与墓地。公寓面向街道和花园，如此一来就可以免受公墓一侧的噪声影响。位于首层的公寓都配置有朝向优良的私人花园（南—西南—东南），可起到过滤街道噪声的作用。每栋建筑都有自己的入口，居民穿过一块小绿地就能来到入口大厅。

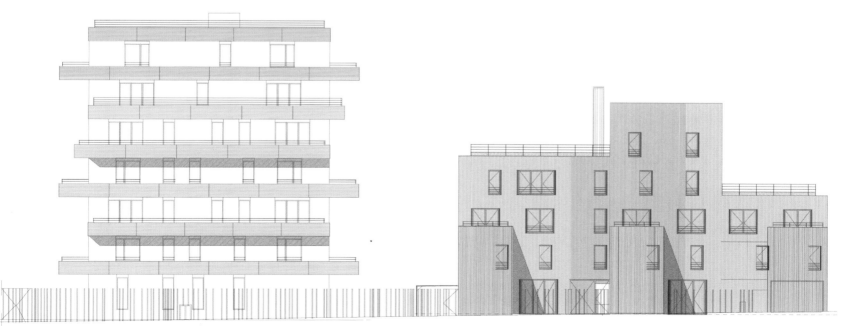

Façade Design
立面设计图

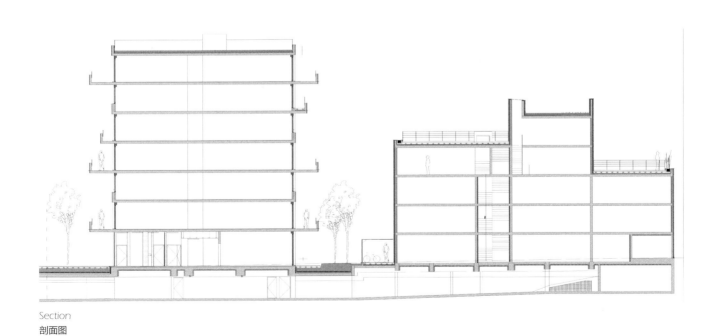

Section
剖面图

In the horizontal black building, each floor houses two or three apartments, which are accessed by a lift or by naturally-lit staircase. These apartments have dual or triple aspects facting outer. The corner south-facing living rooms look out onto the street. The bedrooms face west or east. Despite the proximity of the cemetery, the kitchens and bathrooms benefit from small, high ventilation openings. In this way the apartment are protected from the noise originating from the ring road and views over the cemetery (no windows looks straight out onto the cemetery).

These apartments have generous terraces and balconies that occupy the entire façade of the apartment with access from the living rooms and the bedrooms. Their different floors seem to have been slid one on top of another, giving magnificent views of the sky and giving their residents the feeling that they are living on the top floor — whatever floor their home is actually on.

The black building is a succession of slightly staggered plateau clad in reflective stainless steel. Their different orientations reflect their environment and from the street the cemetery and its greenery can be glimpsed. This brilliant cladding reflects the sunlight and sends shafts of light onto the soffits of the balconies. These plateaux are separated by black concrete walls; sliding wooden shutters form a continuous controlled band. The living rooms all benefit from a corner position and open out onto generous balconies.

水平线条的黑色建筑里，每层都设置有两个或三个单元，配置有电梯和自然光线充足的楼梯。这些公寓都有双面或三面朝外。位于角落朝南的客厅可以直面街道；卧室的方向或朝西或朝东。虽然厨房和浴室靠近公墓，却拥有小型的高通风口。如此一来，所有单元都能免受环路噪声和公墓的影响（没有窗口会直视到墓地）。

这些公寓单元都拥有宽敞的露台或阳台，占据了建筑的整个立面，并且客厅和卧室都连接到阳台。公寓的每个楼层看起来就像是滑动式结构，让每户都能毫无阻碍地观赏天空的壮丽景色，无论是哪层的住户都会感觉自己是住在最高层。

黑色建筑是由一系列略有交错的楼层堆叠而成，表面覆盖有反光的不锈钢外衣。各楼层不同的朝向拥有不同的街景，还能看到公墓的绿树。这层耀眼的外衣能反射阳光，将光线反射到阳台的拱腹上。这些楼层被黑色混凝土墙隔开；滑动的木质百叶窗则是连续的控制带。位于角落的客厅都能连接到宽阔的阳台。

In the wooden building, all the apartments have either a garden (on the ground floor) or large balconies. On the ground floor, the living rooms and bedrooms look out onto the gardens on the street and cemetery sides. In the middle floors, small volumes project out which giving the impression of a smaller volume in terms of the street. The duplex apartments have dual-aspect living rooms benefiting from balconies at both ends. On the top floors, the apartments benefit from having lower level roofs with very generous dimensions.

The wooden building is narrow and hugs the edge of the plot. This volume is completed by small extensions, which house the ground floor bedrooms and form the balconies of the upper floors. The façades are clad in large wooden laths. The building is clad on all four sides so that even when viewed from the cemetery side, the building does not appear impoverished in any way. Openings have been introduced on all sides of the building and all the living rooms have dual aspects giving views of the street and the Honoré-de-Balzac High School. From the other side, the views are of the trees in the cemetery and out beyond the city limits.

Floor Plan 1
楼层平面图 1

木质建筑里的公寓单元都配有花园（位于首层）或大阳台。首层的客厅和卧室可以透过花园看到街景和公墓一侧。在中间楼层，凸出的小型体量让公寓从街道上看起来更为小巧。复式公寓的双面朝外，客厅能从两侧的阳台接收充足的光线。在顶层，因为屋顶较低，顶层的单元能享有更大的面积。

木质建筑较为狭窄，靠近场地的边缘。设计师通过略小的延伸来完成体量的设计，这个延伸则构成了首层的卧室和高层的阳台。建筑的立面覆盖的是大型的木质板条。建筑四面都是木质墙壁，如来一来人们即使从公墓方向看过来也不会觉得很突兀。建筑的各个面都开了许多窗口，所有的客厅都享有双面朝外的位置，以欣赏到街景和 Honoré-de-Balzac 高中的景致；从另一面则能看到公墓的绿树，甚至是城市的边缘。

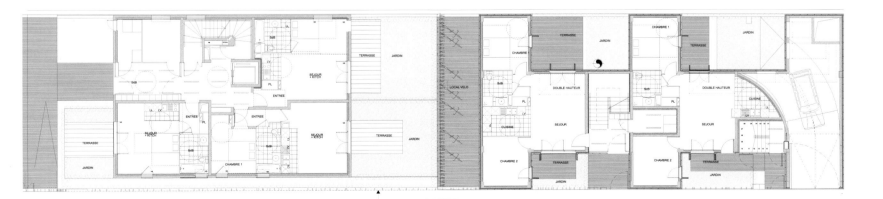

Floor Plan 2
楼层平面图 2

MATERIAL 材料
FAÇADE 立面
STRUCTURE 结构
LAYOUT 布局

Cemetery Road Apartment Building, Sheffield, UK

英国谢菲尔德陵园路住宅楼

Architect: Project Orange architects and designers
Client: Neaversons
Location: Sheffield, UK
Site Area: 3,600 m²
Gross Floor Area: 2,800 m²
Floors: 4

设计公司：Project Orange 建筑设计事务所
客户：Neaversons
地点：英国谢菲尔德市
占地面积：3 600 平方米
建筑面积：2 800 平方米
层数：4

The project is a 3,600 m² new build mixed-use scheme situated one mile southwest of Sheffield, 2 km away from city centre. The project is the second undertaken for the client of the successful Sinclair Building in the city. The site borders the local conservation area to the south and east and is located within an established inner suburb, adjacent to the Sheffield General Cemetery.

本项目是一栋占地 3 600 平方米的新综合大楼，距离谢菲尔德市中心的西南部 2 km，是本案客户继辛克莱大厦后在本城成功建设的第二个项目。场地扩展了保护区的南部和东部，位于已建成的内郊区，毗邻谢菲尔德将军坟。

FAÇADE

The elevations facing the conservation area are clad in locally sourced traditional Yorkshire stone, a material that was widely used throughout the city throughout the nineteenth and early twentieth century. We have juxtaposed the smooth ashlar version of the stone with its rough hewn riven faced counterpart and use these in a composition that reveals the contemporary nature of the new buildings. Within the protective armature of these stone façades, the courtyard elevations are intended to be lightweight and informal reflecting the more private nature of this outdoor room.

立面

面对保护区的立面覆层采用约克郡当地的传统石材，这是在19世纪和20世纪初时城市里广泛使用的材料。设计师将石材并列成平滑的方石模式，粗糙的一面面对中庭。这样的组合展示出建筑的当代特性。有了这些石材立面的保护，中庭立面会更显轻质，不经意地勾勒出户外空间的私密性质。

Site Plan
总平面图

Elevation from Cemetery
墓园方向立面图

Elevation from Cemetery Road
陵园路立面图

Front Dorsal View Of Elevation
前部背视图

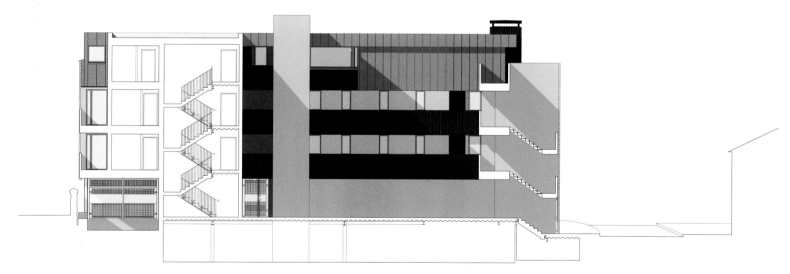

This high-density project proposes nine townhouses flanking a raised courtyard garden, all above a shared semi-basement car park, located to take advantage of the sloping site. Three commercial units at ground level with six apartments and an 180 m² penthouse above are located within a four-storey block that fronts Cemetery Road and provides the public face of the scheme.

本案是一个高密度项目，高起的中央花园两侧分布着 9 座联排别墅，利用坡地的优势共享一个半地下停车场。每栋建筑首层是 3 个商店，上面的 4 层分别是 6 间公寓和一间 180 平方米的复式套房，既面对陵园路，又是建筑向公共开放的正面。

The internal arrangements of the two blocks of houses are configured to maximise privacy between neighbours with the principal habitable rooms of the easterly block facing the staircases and ancillary accommodation to the rear across the courtyard. Both house types are designed with dynamic cross sections and make a generous provision of balcony and roof terrace areas.

两栋建筑之间的内部结构经过特别配置，最大程度地保护邻居之间的隐私。例如，朝向东方的建筑的主居室就面向楼梯，附属设施则设置在穿过中庭的后方。户型的设计都注重动感的结构，并附带有宽敞的阳台和屋顶露台区。

East Elevation
东侧立面图

West Elevation
西侧立面图

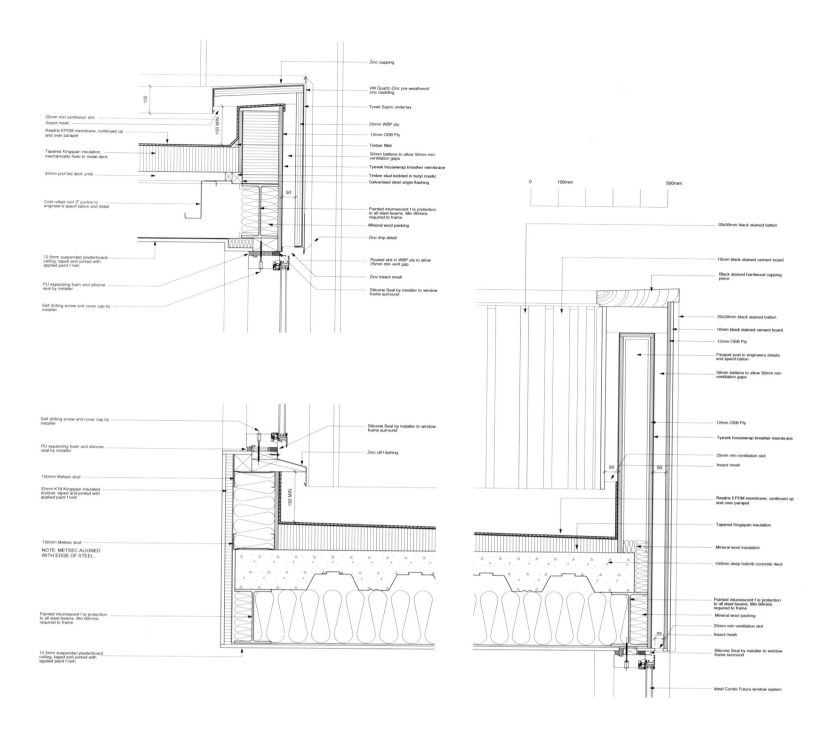

25mm min ventilation slot
Insect mesh

Resitrix EPDM membrane, continued up
and over parapet

Tapered Kingspan insulation,
mechanically fixed to metal deck

32mm profiled deck units

Cold rolled roof 'Z' purlins to
engineer's specification and detail

12.5mm suspended plasterboard
ceiling, taped and jointed with
applied paint finish

PU expanding foam and silicone
seal by installer

Self drilling screw and cover cap by
installer

100

150 MIN

50

Zinc capping

VM Quartz-Zinc pre-weathered
zinc cladding

Tyvek Supro underlay

25mm WBP ply

12mm OSB Ply

Timber fillet

50mm battens to allow 50mm min
ventilation gaps

Tyevek housewrap breather membrane

Timber stud bedded in butyl mastic

Galvanised steel angle flashing

Painted intumescent fire protection
to all steel beams. Min 60mins
required to frame

Mineral wool packing

Zinc drip detail

Routed slot in WBP ply to allow
25mm min vent gap

Zinc insect mesh

Silicone Seal by installer to window
frame surround

Self drilling screw and cover cap by
installer

PU expanding foam and silicone
seal by installer

150mm Metsec stud

25mm K18 Kingspan insulated
dryliner, taped and jointed with
applied paint finish

150mm Metsec stud
NOTE: METSEC ALIGNED
WITH EDGE OF STEEL

Painted intumescent fire protection
to all steel beams. Min 60mins
required to frame.

12.5mm suspended plasterboard
ceiling, taped and jointed with
applied paint finish

150 MIN

Silicone Seal by installer to window
frame surround

Zinc cill flashing

0   100mm   500mm

20x30mm black stained batten

10mm black stained cement board

Black stained hardwood capping
piece

20x30mm black stained batten

10mm black stained cement board

12mm OSB Ply

Parapet post to engineers details
and specification

50mm battens to allow 50mm min
ventilation gaps

12mm OSB Ply

Tyevek housewrap breather membrane

25mm min ventilation slot

Insect mesh

Resitrix EPDM membrane, continued up
and over parapet

Tapered Kingspan insulation

Mineral wool insulation

150mm deep holorib concrete deck

Painted intumescent fire protection
to all steel beams. Min 60mins
required to frame

Mineral wool packing

25mm min ventilation slot

Insect mesh

Silicone Seal by installer to window
frame surround

Ideal Combi Futura window system

50   50

50

25

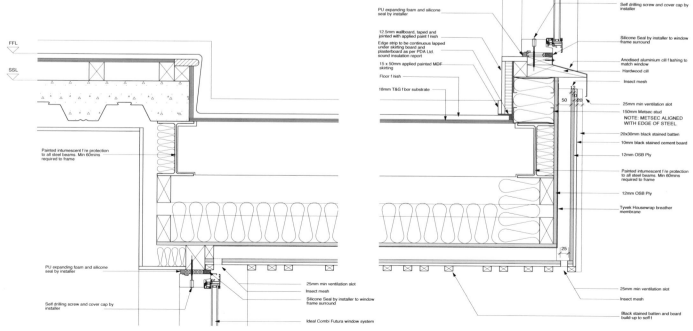

FFL

SSL

Painted intumescent fire protection
to all steel beams. Min 60mins
required to frame

12.5mm suspended plasterboard
ceiling, taped and jointed with
applied paint finish

PU expanding foam and silicone
seal by installer

Self drilling screw and cover cap by
installer

25mm min ventilation slot

Insect mesh

Silicone Seal by installer to window
frame surround

Ideal Combi Futura window system

PU expanding foam and silicone
seal by installer

12.5mm wallboard, taped and
jointed with applied paint finish

Edge strip to be continuous lapped
under skirting board and
plasterboard as per PDA Ltd.
sound insulation report

15 x 50mm applied painted MDF
skirting

Floor finish

18mm T&G floor substrate

Self drilling screw and cover cap by
installer

Silicone Seal by installer to window
frame surround

Anodised aluminium cill flashing to
match window

Hardwood cill

Insect mesh

25mm min ventilation slot

150mm Metsec stud
NOTE: METSEC ALIGNED
WITH EDGE OF STEEL

20x30mm black stained batten

10mm black stained cement board

12mm OSB Ply

Painted intumescent fire protection
to all steel beams. Min 60mins
required to frame

12mm OSB Ply

Tyvek Housewrap breather
membrane

25mm min ventilation slot

Insect mesh

Black stained batten and board
build-up to soffit

50

25

50   20

Working Detail of Western Townhouse Façade
西侧联排别墅立面细节图

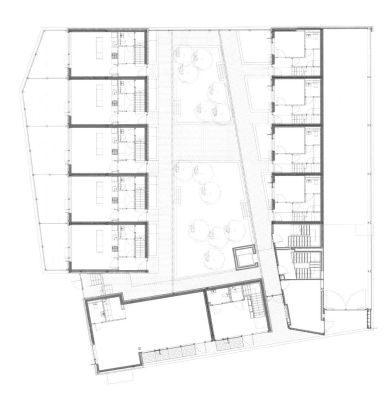

Ground Floor Plan
首层平面图

1st Floor Plan
2 层平面图

2nd Floor Plan
3 层平面图

3rd Floor Plan
4 层平面图

COLOR 颜色

STRUCTURE 结构

WINDOW 窗

FAÇADE 立面

Surpass Court, Shanghai, China

中国上海永嘉庭

Architect: logon
Location: Shanghai, China
Site Area: 3,744 m²
Gross Floor Area: 7,422 m²

设计公司：罗昂建筑设计咨询有限公司
地点：中国上海市
占地面积：3 744 平方米
建筑面积：7 422 平方米

Surpass Court is located on Yongjia Road in the former French Concession, Shanghai, which was under French control from 1849 to 1943. The surroundings are mainly low rise Shanghai typical lane houses (called Lilong) and French style villas, dating mainly from the 1920 to 1930s. Surpass Court is an adaptive reuse project of utilizing outdated aviation research institute buildings by turning them into a new urban lifestyle center. Now, as a commercial, office and F&B destination, it is a good example that how through design, reutilizing a former industrial area not only can become a public space that brings people together, but can also have the ability to rejuvenate its' surroundings, generating a new lifestyle for the community.

永嘉庭坐落于上海原法租界的永嘉路，这里在1849年到1943年间是法国的占领区。周边主要是上海典型的低层里弄和19世纪二三十年代的法式别墅。永嘉庭是一个改建项目——将原航天研究所的办公楼改造为都市时尚中心。如今，这个集商业、办公、餐饮等各种功能为一体的案例展示了如何通过设计将老楼铸就为积聚人气的公共空间，并以其全新的生活方式带动所在的社区焕发勃勃生机。

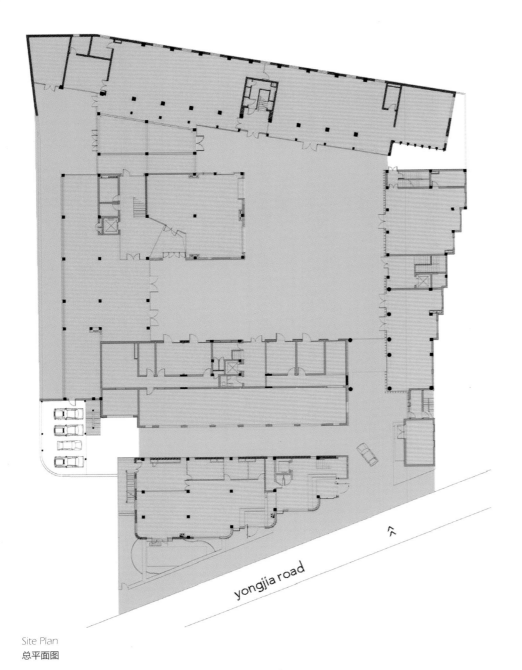

Site Plan
总平面图

The project covers an area of 3,744 m² and offers 7,442 m² of floor area. The project is set back from the road, facing Yongjia Road only with one out of 9 façades within street view, while the other façades encircle an inner courtyard set back from the street, accessible only by one entrance gate. The secluded quality of the courtyard speaks the true potential of the site where people gather together day and night for festivals and events, providing space for activities in a central location without disturbing the immediate neighbors. The street facing commercial space is positioned for a variety of restaurants and bars, immediately giving passersby an indication what to find inside the courtyard, sparking curiosity for people to investigate and enter the site. The main challenges for the design team was how to create an identity for the courtyard, and in addition to that, how to work with the existing architecture, not against it.

Shanghai is a city that comes alive at night; Surpass Court is no exception taking advantage of its modest façades with soft lighting which illuminates the building's exterior attracting restaurant and bar goers alike. This has triggered a flow-on effect where now new F&B outlets and other cultural industries leverage the success of Surpass Court rejuvenating its' surrounding area.

该项目占地面积 3 744 平方米，建筑面积 7 422 平方米。庭院并不临街，所有楼宇的 9 个立面只有一个是面向永嘉路的，其余立面围合成一个远离街面的庭院，出入口且只有一个。与世隔绝的静谧是这个庭院的潜力所在：人们在节假日的日日夜夜集聚在这里举行活动，却并不会惊扰到周边的邻里。临街商业空间定位为各种酒吧和餐厅，旨在给予路人一种暗示，激发他们走进来探寻庭院中的一切。设计上主要的挑战在于如何为这个庭院定位，以及如何利用现有的建筑，而不是与它们对立。

上海是一个不夜城，永嘉庭也无一例外地利用了柔和的灯光照亮了它时尚的立面，吸引着夜生活的拥戴者们前来。永嘉庭成功的群聚效应还吸引着众多休闲餐饮以及文化场所来到这个区域，并不断改善着它的周边环境。

FEATURE 特点分析

STRUCTURE

The white façade reflects an elegant style. A little shimmer at night will light up the whole building particularly and conspicuously, which attracts tourists effectively.

结构

白色的立面尽显雅致的格调，在夜间，仅需些许微光就可使建筑格外显目，从而吸引了更多的游客。

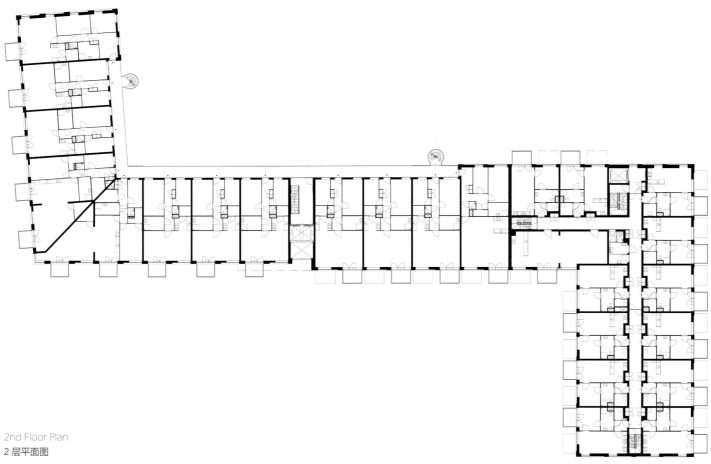

2nd Floor Plan
2 层平面图

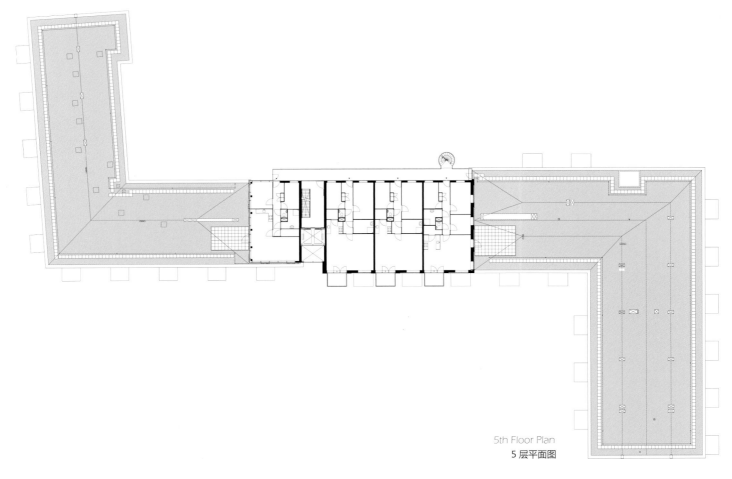

5th Floor Plan
5 层平面图

ACKNOWLEDGEMENTS

鸣谢

NARCH (Joan Ramon Pascuets, Monica Mosset)
Block 20 Dwellings and Commercial, Barcelona, Spain

MVRDV
Westerdok Apartment Building, Amsterdam, the Netherlands

Project Orange architects and designers
Cemetery Road Apartment Building, Sheffield, UK

Rogers Stirk Harbour + Partners
One Hyde Park, London, UK

Neil M. Denari Architects
High Line 23, New York, USA

Spine Architects
La Taille Vent, Hamburg, Germany

Cino Zucchi Architetti
Ex Rossi Catelli Area Residential Building, Italy

MDR Architectes
Building 29 Living Units and Shops, Hérault, France

OFIS
Basket Apartments in Paris, France

Casanova+Hernandez Architects
Ginkgo Project

Make Architects
10 Weymouth Street, London, UK
Grosvenor Waterside, London, UK

Zucchi & Partners (Cino Zucchi, Nicola Bianchi, Andrea Viganò)
Darsena Lot 4 Apartment Building on the Harbor, Ravenna , Italy

Arons en Gelauff architecten
De Entrée, Alkmaar, the Netherlands

BFTA Mimarlık Ltd.
Miracle Residence, Istanbul, Turkey

Open Building Research S.r.l.
Milanofiori Residential Complex, Milano, Italy

PERIPHERIQUES Architectes
Fremicourt, Paris, France

Sheppard Robson
Grosvenor Waterside, London, UK

Perkins Eastman
303 East 33rd Street, New York, USA

Rojkind Arquitectos
High Park, Monterrey, Mexico

JDS
Iceberg in the Aarhus Docklands, Denmark

CEBRA
Iceberg in the Aarhus Docklands, Denmark

Louis Paillard
Iceberg in the Aarhus Docklands, Denmark

SeARCH
Iceberg in the Aarhus Docklands, Denmark

3XN
Green School Stockholm, Sweden

Studio Daniel Libeskind
Fiera Milano, Milan, Italy

ECDM
RATP Apartment, Paris, France

Nakae Architects
NE Apartment, Tokyo, Japan

Akiyoshi Takagi Architects
NE Apartment, Tokyo, Japan

Francis -Jones Morehen Thorp
Capella Apartments, Sydney, Australia
Little Bay, Australia

Nabil Gholam Architecture
Beirut Marina & Town Quay, Beirut, Lebanon

Philippe Gazeau Architecte
Logements La Courrouze, Rennes, France

ORMS Architecture Design
"The Writers", London, UK

Forum Architects
Shelford Road Apartments, Singapore

Dick van Gameren Architecten
Block A, Amsterdam, the Netherlands

Metaphorm Architects
Brandon Street Affordable Housing, London, UK

C. F. Møller Architects
The Fuglsang Lake Centre, Herning, Denmark

Agence Christophe Rousselle
Habitations Rue Pierre-Rebière, Paris, France

Agence Nicolas Laisné
Habitations Rue Pierre-Rebière, Paris, France

J. Mayer H. Architects
Schlump ONE

logon
Surpass Court, Shanghai, China

Ohno JAPAN
NE Apartment, Tokyo, Japan

Patterns + MSA
Jujuy Redux, Rosario, Argentina

Steven Holl Architects
Beirut Marina & Town Quay, Beirut, Lebanon

Special thanks to the architects above for their high-quality projects and nice supports, Inquiries or suggestions are welcome at any time to:
news@hkaspress.com

特别鸣谢以上设计公司的一贯支持，为本书提供优质的作品，如有任何问题或建议请联系：
news@hkaspress.com

图书在版编目（ＣＩＰ）数据

突破风格与复制Ⅱ．下／香港建筑科学出版社编．
—天津：天津大学出版社，2013.6
ISBN 978-7-5618-4705-3

Ⅰ．①突…Ⅱ．①广…Ⅲ．①住宅 - 建筑设计 - 世界
Ⅳ．① TU241

中国版本图书馆 CIP 数据核字 (2013) 第 120422 号

责任编辑　郝永丽
装帧设计　黄　丹　王双玲
文字编辑　罗小敏　纪文明　黄夏炎　黄乐琪
流程指导　陈小丽
策划指导　高雪梅

突破风格与复制Ⅱ（下）

出版发行　天津大学出版社
出 版 人　杨欢
地　　址　天津市卫津路 92 号天津大学内（邮编：300072）
电　　话　发行部 022-27403647
网　　址　publish.tju.edu.cn
印　　刷　利丰雅高印刷（深圳）有限公司
经　　销　全国各地新华书店
开　　本　245mm×325mm 1/16
印　　张　46
字　　数　657 千字
版　　次　2013 年 9 月第 1 版
印　　次　2013 年 9 月第 1 次
定　　价　736.00 元